Fenwick Thomas

Elementary and Practical Treatise on Subterraneous Surveying

and the Magnetic Variation of the Needle by Thomas Fenwick;

Fenwick Thomas

Elementary and Practical Treatise on Subterraneous Surveying
and the Magnetic Variation of the Needle by Thomas Fenwick;

ISBN/EAN: 9783744738712

Printed in Europe, USA, Canada, Australia, Japan

Cover: Foto ©berggeist007 / pixelio.de

More available books at **www.hansebooks.com**

ELEMENTARY AND PRACTICAL TREATISE

ON

SUBTERRANEOUS SURVEYING,

AND THE

MAGNETIC VARIATION OF THE NEEDLE.

BY THOMAS FENWICK,
Colliery Viewer and Surveyor of Mines.

ALSO

THE METHOD OF CONDUCTING

SUBTERRANEOUS SURVEYS WITHOUT THE USE OF THE MAGNETIC NEEDLE,

And other Modern Improvements.

BY THOMAS BAKER, C.E.,
Author of "Railway Engineering," "Theodolite Surveying, Levelling," &c., in Nesbit's "Surveying, Rudimentary, Land and Engineering Surveying, Statics and Dynamics, Elements of Mechanism, Mensuration, Integration," in Weale's Series, &c.

THE THIRD EDITION.

LONDON:
JOHN WEALE, 59, HIGH HOLBORN.
1861.

PREFACE.

The mineral wealth of this kingdom had become of such great importance, about half a century ago, as to induce *Mr. T. Fenwick, of Dipton, in the County of Durham*, to compose a Treatise on Subterraneous Surveying (which forms the basis of the present Work) for the use and instruction of young men designed for the profession of mining agents and surveyors, usually called colliery viewers: much more, then, is such a treatise now necessary, as these mineral productions have, up to the present time, been more than quadrupled in value; and by the more general diffusion of mathematical, philosophical, and mechanical science, the working of mines has been conducted with greater skill and precision for the full development of their vast wealth.

The general use of the magnetic needle in subterraneous surveys has been found to be a great source of error, on account of ferruginous substances (which exist in almost all mines) attracting the needle, and causing it to give erroneous indications; whence, in general, old surveys are found to be extremely defective. Indeed, *Mr. Fenwick* himself was so sensible of this deficiency of the needle, that he proposed, in the Second Edition of his Work, about forty years ago, to dispense with its general use; though he still proposed to use it, at the first departure, or commencement of the survey, from the top to the bottom of the shaft of the mine.

This Edition of the Work contains, in a small compass, the essentials of Subterraneous Surveying in all its branches, both with and without the use of the magnetic needle; and to make it still more useful to that class of men for whom it is chiefly intended to convey information, there are added a great number of explanatory figures and examples.

Part I. contains the method of surveying, with the use of the magnetic needle, without attending to its variation, as being more readily intelligible to beginners; and the magnetic bearings being, at the same time, at once adapted to the use of the Traverse Tables. This part is arranged after *Mr. Fenwick's* plan (whose method and examples are still retained), in the following order:—

1. Geometrical problems.

2. Theorems, and the methods of conducting subterraneous surveys.

3. Of determining the magnitude of angles.

4. Of determining bearings, and reducing angles to the bearings which they form with the magnetic meridian, with a rule and examples.

5. The method of reversing bearings.

6. Of reducing bearings to the angles they form with the magnetic meridian, with rules and examples; and the manner of finding the magnitude of the angle that two bearings form with each other.

7. The method of reducing bearings and distances to the northing or southing, and easting or westing, they contain, by the Traverse Table, with a rule and examples.

8. The manner of surveying subterraneous excavations with the form of the survey-book.

9. The method of taking back sights.

Part II.—In this part, which treats extensively on conducting subterraneous surveys, without the use of the magnetic needle, Mr. Fenwick's examples are in several cases retained, with full directions for adapting them to the new method (they being already adapted to the use of the Traverse Table), which will constitute a useful exercise for the student in transferring the angles from their magnetic bearings to the angles which one line makes with the preceding one, as taken by the theodolite. This part has the following arrangement:—

1. Mr. Fenwick's method of subterraneous surveying, without the use of the needle, except at the first departure or commencement of the survey.

2. Mr. Baker's method of commencing the survey by suspending two weights down the shaft in the direction

of the first headway, and marking the same direction on the surface; and afterwards conducting the survey with the theodolite, without the use of the needle.

3. Mr. Beauland's method of making the commencement of the survey by the help of a transit instrument, not using the needle, as in Baker's method.

4. Plotting and protracting surveys in various ways.

5. Of reducing the bearings and distances of a survey into one common bearing and distance, or any number of bearings and distances fewer than those that compose the survey, whether the angles be taken with the needle, or the theodolite independent of the needle.

6. The method of plotting on the surface in various ways.

7. The method of making the survey where the excavation inclines from the horizon.

8. A promiscuous collection of practical examples, some of which relate to tunnelling.

PART III. contains subterraneous surveys, under the necessary attention to the magnetic variation of the needle. As the magnetic meridian has been found to be in a state of variation from the true meridian for upwards of 300 years, and still continues to vary, therefore surveys made by the circumferentor, or any other instrument under magnetic influence, must vary accordingly as that meridian varies. For instance, suppose the bearing of any one known object to have been taken from a given point by the magnetic meridian in the year 1700, and recorded; and if the bearing of the same object be now retaken by the magnetic meridian from the same given point, these two bearings will be found, on comparison, to differ about 14°, the magnetic meridian having in that time changed thus far in its direction (see table, p. 96). It is also well known to directors of mines that the plans of their excavations, on examination, are always found to be erroneous,—some even to a great extent. This frequently misleads the miner, adding expense to his subterraneous pursuits, and the cause of such errors originates through his inattention to the variation of the needle in the plotting from time to time of his surveys.

This part, therefore, shows the method of rectifying the bearings of old surveys, in order to connect them with those made by the scientifically correct method laid down in the second part of this work.

The third part is thus arranged:—

1. Axioms and observations.

2. The method of finding the true and invariable meridian.

3. To determine the variation of the needle of the circumferentor or other instrument used in surveying.

4. To reduce bearings taken by an instrument, the needle of which has any known variation, to bearings with the true meridian, with rules and examples.

5. To reduce bearings from one magnetic meridian to bearings with any other magnetic meridian, with rules and examples.

6. To find the kind of meridian by which a plan has been constructed, with rules and examples.

7. On planning surveys, and finding the magnitude of an error in plotting, caused by inattention to the magnetic variation, with examples.

8. On running bearings on the surface by the circumferentor or theodolite without error.

9. To determine the antiquity of a plan by its delineated meridian.

10. On recording bearings.

11. The Traverse Tables, with examples of their use.

12. An expeditious method of calculating the produce of coal strata of any given thickness, with examples.

13. Concluding examples in mining surveying.

Having now described the plan, and enumerated the heads of this publication, I must leave it to practical colliery viewers of scientific skill to judge of its merits and utility, in its present improved form; and I trust, from my own practical experience in surveys of almost every kind during the last forty years, that the difficulties and intricacies of such a work will, to candid and liberal minds, be sufficiently obviated.

T. BAKER.

CONTENTS.

PART I.

PART II.

PART III.

EXPLANATION

OF

TERMS AND EXPRESSIONS IN THIS WORK.

Bearing to the right or left of a meridian. A line is said to bear on the right or left of the north or south meridian, when it is to the right or left of a person, whose face is turned towards the north or south.

Bearing on different sides of a meridian. Two lines are said to bear on different sides of a meridian, when the one bears on the east side, and the other on the west side thereof.

A Bord is an excavation in a seam of coal driven in a direction across its fibres.

A Drift is a narrow excavation driven in any direction in coal or stone.

A Headway is an excavation in a seam of coal driven in the direction of its fibres.

Different Meridians. When one line bears in a given direction with the north meridian, and another bears in a given direction with the south meridian, those lines are called bearing with different meridians. Also, when one line bears on the east side of the north meridian, and another on the west side of the south meridian, those lines are said to bear on different sides of different meridians, and *vice versâ.*

TREATISE

ON

SUBTERRANEOUS SURVEYING,

ETC.

PART I.

GEOMETRICAL PROBLEMS.

1.—*To divide a given line* AB *into equal parts.*

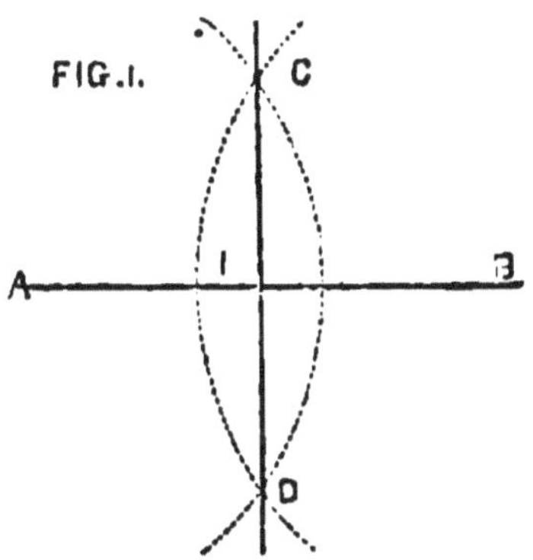

WITH any distance greater than half AB, and one foot of the compasses on A and B, describe two arches cutting each other in C and D; through the intersecting points CD draw a line CD, which will cut AB in I into equal parts.

2.—*To draw a line parallel to a given line* CD, *to pass through any assigned point* A.

From the given point A take the nearest distance to the given line CD; with that distance, and one foot of the compasses, any where towards C describe an arch O; through A draw a line AB, just to touch the arch O in O; and the line AB will be the parallel required.

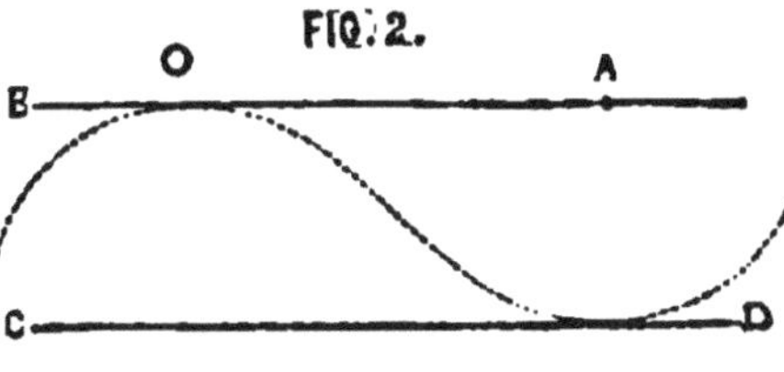

3.—*To raise a perpendicular from a given point* P *in a given line* AB.

From the given point P describe the arch FD; take PF, and set from F to C, and from C to D; then with any convenient distance from C and D describe the arches O, and through their point of intersection from the point P draw the line PO, the perpendicular required.

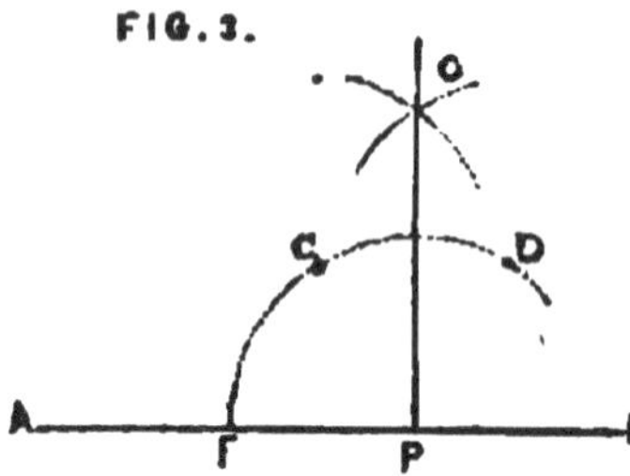

4.—*To raise a perpendicular from a given point* A, *at the end of a given line* AB.

Set one foot of the compasses on A, and extend the other to any point C, above the line AB; on the centre C describe the semicircle FAP, to cut AB in F; draw FC cutting the semicircle in P; then draw AP, which will be perpendicular to AB.

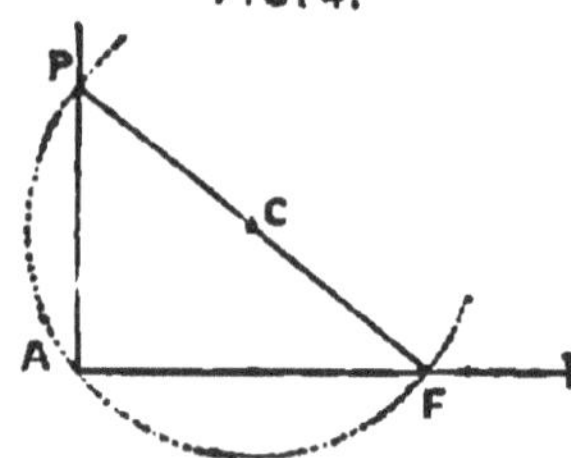

5.—*From a given point* P *to let fall a perpendicular upon a given line* AB.

On the given point P as a centre, describe the arch EF to cut AB in E and F; with any convenient distance, and one foot of the compasses on E and F, describe two arches to cut each other in I; through P and I draw PI, which is perpendicular to AB.

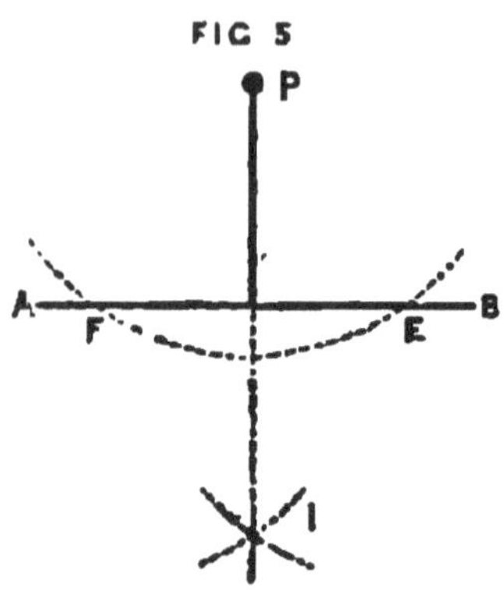

6.—*To make an angle* ABC *equal to a given angle* CDE.

With any convenient extent of the compasses, and one foot on D, draw the arch FG; equal to the measure of the given angle D draw a line BC, and with the distance DF describe the arch HI; then make the arch HI equal to the arch FG, and through I draw the line BA, forming the angle;—so the angle ABC is equal to the angle CDE.

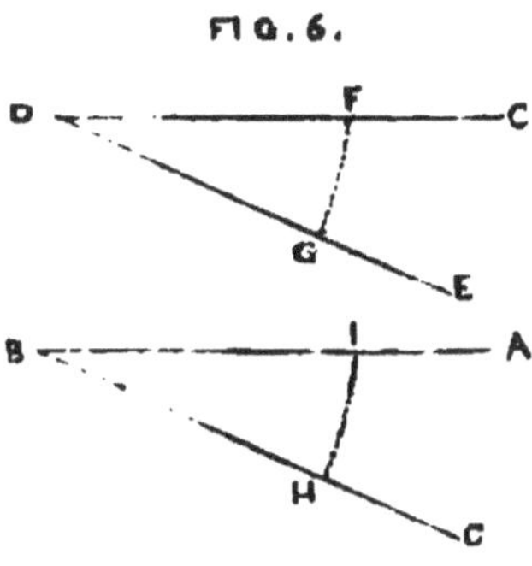

7.—*To lay down an angle* FDG *equal to any determined number of degrees, which suppose* 35°

Draw the line DF at pleasure, and with 60° off the scale of chords describe the arch EH on the centre D; from the same chords take 35° (the quantity of the angle), and lay upon the arch from E to H, through which from D draw the line DG, and the angle FDG will contain just 35°.

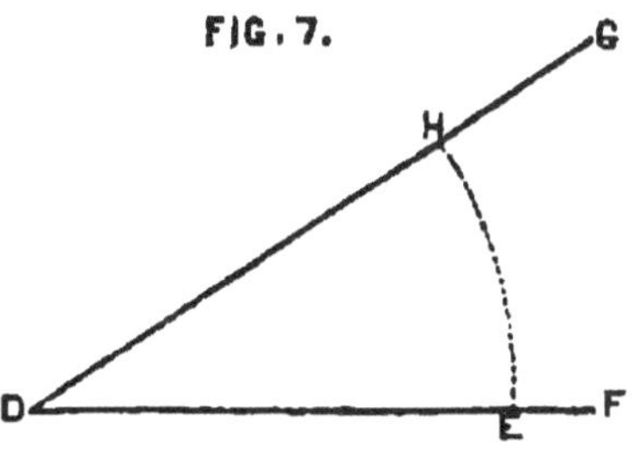

8.—*To determine the number of degrees contained in any angle, suppose angle* FDG.

With 60°, taken from the scale of chords, describe the arch EH; then extend the compasses from E to H, and observe, on the same line of chords, what number of degrees the extension measures,—which will be the measure of the angle EDH.

Or, apply the centre of the protractor to the angular point D, and bring its straight edge upon the line DF, and the degree the other line cuts on the divided arch is the measure of the angle.

THEOREMS.

1. Every right angle, as ACB, contains 90 degrees or equal parts.

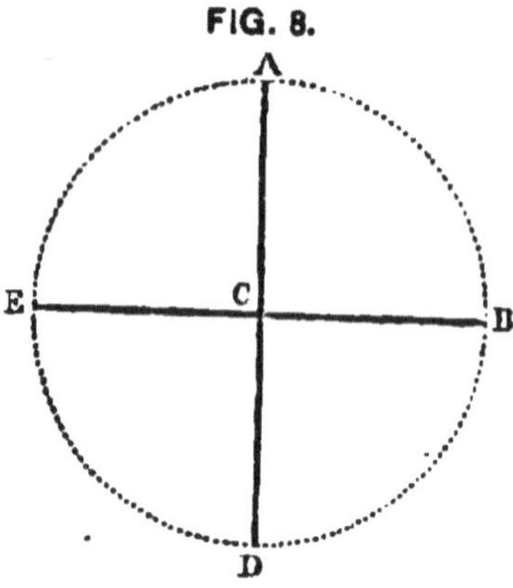

FIG. 8.

2. Every circle ABDE, is supposed to have its periphery divided into, or to contain, 360 equal parts, called degrees, —and those degrees are divided into 60 equal parts, called minutes,—and each minute is divided again into 60 equal parts, called seconds, &c.

3. Every circle AD, contains four right angles, at angles ACB, BCD, DCE, and ECA, which, from theorem 1, must contain 90° each.

4. Every semicircle EAB, contains two right angles, as angles ECA and ACB, which, from theorem 1, must contain 90° each.

Draw the diameter AD, which will divide the circle EABD into two equal parts EAB and EDB, each containing a semicircle, or 180°; if, therefore, a line AC be drawn perpendicular to EB from the centre C, it will divide the semicircle EAB into two equal parts, making two right angles ECA, ACB.

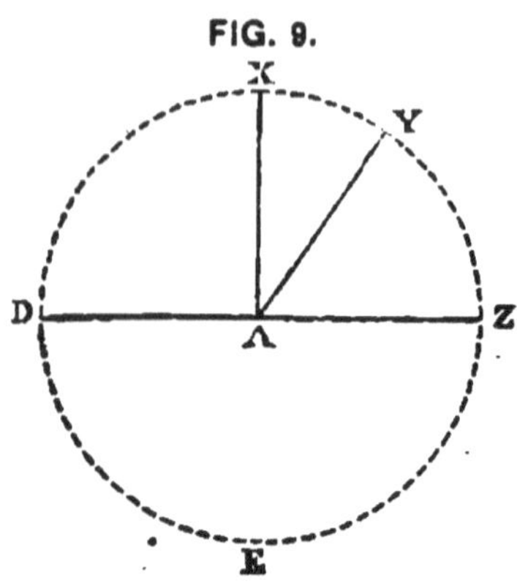

FIG. 9.

5. If any right line AY stands upon another right line DZ, it will make therewith two right angles, or two angles whose sum is equal to two right angles.—(Euc. b. 1, p. 13.) If a line AY, be drawn from any part Y of the circumference to A, it will

divide the semicircle DXZ into two unequal parts, making the angles DAY, YAZ, unequal; but these two angles are equal to a semicircle, or two right angles.

6. If two right lines IL, KM, intersect each other, the opposite angles A and C, as also B and D, are equal; that is, the angle A = the angle C, and the angle B = the angle D.—(Euc. b. 1, p. 15.)

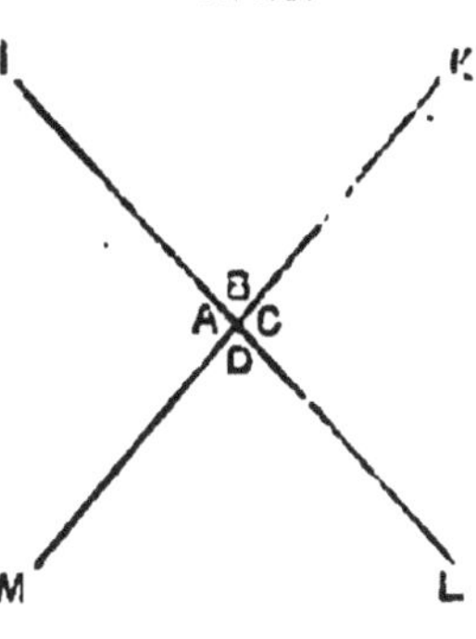

7. If a right line OR, cuts two parallel right lines NP and SQ, the alternate angles NaR, QbO, are equal, and consequently the lines parallel.—(Euc. b. 1, p. 29.)

8. If any side of a right-lined triangle be continued, see fig. 12, the external angle is equal to the sum of the two opposite internal ones.—(Euc. b. 1, p. 32.)

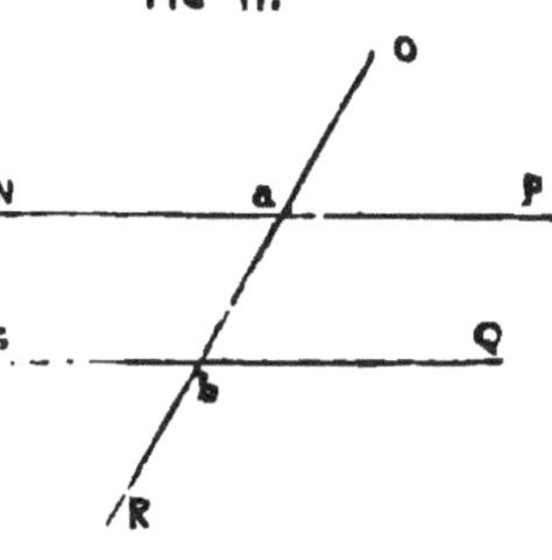

Let UST be the given triangle; then the ∠ STZ is = ∠ SUT + ∠ UST, = the sum of the opposite internal angles.

9. The three angles of any triangle are together equal to two right angles, or 180°.—(Euc. b. 1, p. 32.) See fig. 12.

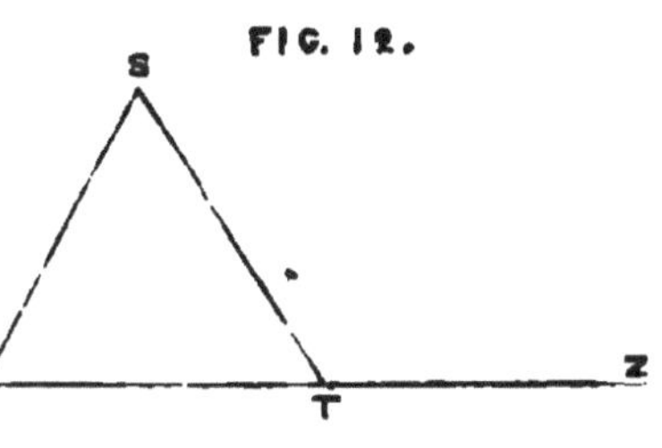

In the triangle STU, the ∠ STU + ∠ TSU + ∠ SUT = 180°, or two right angles.

10. The sides of similar triangles are proportional, and the angles subtended by proportional or equal sides are equal.—(Euc. b. 6, p. 45.)

11. In any four-sided right-lined figure, called a square parallelogram, rhombus, trapezium, &c., the sum of the

four angles is equal to four right angles, or 360°.—(Euc. b. 1, p. 32.)

12. The sum of all the angles of any right-lined figure (though it contain never so many sides) is equal to double as many right angles, abating four, as there are sides in the figure.—(Euc. b. 1, p. 32.)

13. In right-lined triangles, equal sides subtend equal angles (Euc. b. 1, p. 5). The greatest side subtends the greatest angle (Euc. b. 1, p. 19), and the least side subtends the least angle.

14. An angle in a semicircle is a right angle; or if two lines, as TR, SR, be drawn from T and S (the ends of the diameter) to R in the circumference, they will form a right angle TRS.—(Euc. b. 3, p. 31.)

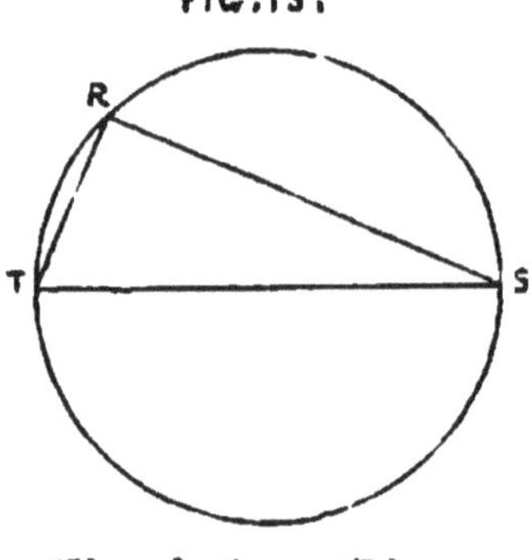

15. In any right-angled triangle, the square of the hypothenuse (or longest side) is equal to the sum of the squares of the other two sides or legs.—(Euc. b. 1, p. 47.)

16. The compass is divided into four cardinal points, called north, south, east, and west; the two first, north and south, are formed where the meridian cuts the horizon,—and the other two, east and west, are each 90 degrees distant from the points north and south; therefore they divide a circle into four equal parts of 90 degrees each.

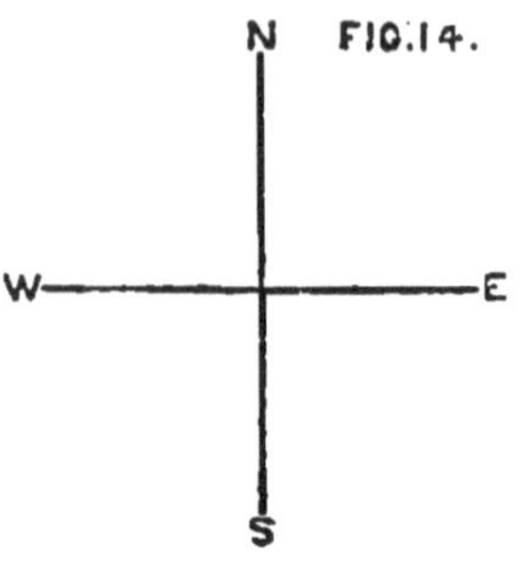

17. When the face is turned to the north N, the right hand is towards the east E, and the left hand towards the west W; and when the face is turned towards the south S, the right hand is towards the west W, and the left hand towards the east E.

18. The magnetic meridian is that line in which the magnetic needle of the compass settles; and every particular place on the earth has its respective magnetic meridian.

19. The magnetic needle is here assumed to retain its parallelism in every situation within the limits of a subterraneous survey.

If in the situation A, a magnetic needle is placed, and is found to settle in the direction of *ab*, — if the same needle is removed to B or C, it will settle itself in the direction of *cd* and *ef*, both parallel to *ab*. But the magnetic meridian of places very distant from each other will not be parallel; for the magnetic meridian of London will vary a few degrees from its parallelism with that of Edinburgh. The magnetic needle has a small diurnal variation, being greatest about noon, also a small annual variation, which seldom exceeds a few minutes of a degree.

FIG. 15.

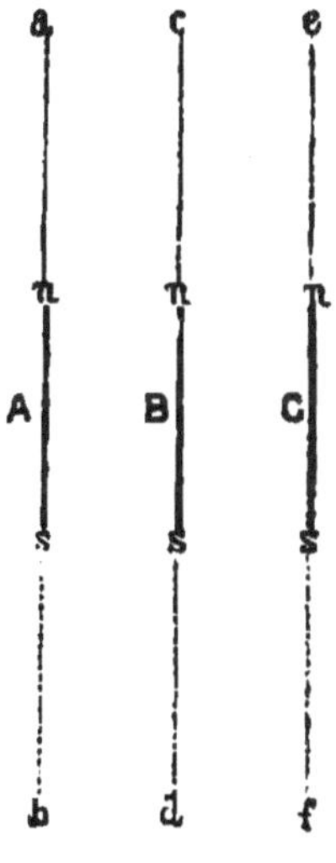

Part First of this work consists of the manner of surveying under-ground, without attending to the magnetic variation of the needle,—with several easy and expeditious modes of plotting the same.

The instruments used in subterraneous surveying are the circumferentor, the theodolite, Gunter's chain, in the coal mines, which contains 100 links. In the lead mines, a chain, divided into 100 feet, is now frequently used instead of Gunter's chain.

The manner of conducting a subterraneous survey by the magnetic needle.

(1.) Place the circumferentor, or instrument used, where

the survey is intended to commence; then let a person go forward in the direction of the line to be surveyed, with a lighted candle in his hand, to the utmost distance his light can be seen through the sights of the instrument; its bearing then is taken by the circumferentor (the manner of taking bearings will be shown hereafter), and noted down in the survey book; proceed then to take the distance of the light or object from the instrument; remove the instrument, and let a person stand on the exact spot where it stood, holding in his hand one end of the chain, while another, going towards the object, holds the other end, together with a lighted candle, in the same hand; then being directed by the former until that hand which holds the candle and the chain is in a direct line with the object or light whose bearing was taken, there mark the first chain; then he that stood where the instrument was placed comes forward to the mark at the end of the first chain, the other advancing another chain forward, with the candle and chain in the same hand, directed as before, there mark the second chain, — so proceeding in the same kind of way until the distance of the object is determined, which being noted down in chains and links in the survey book, opposite to the bearing, then the first bearing and distance is completed:—Fix the instrument again where the light, as an object, stood, or at the termination of the preceding bearing and distance, and take the second bearing, by directing the person to go forward as before, so far as his light can be seen, or at any shorter convenient distance, and proceed as before until the whole is completed.

There should not be fewer than five people employed in such surveys, to carry forward the work with expedition, —viz., one to carry forward the survey, and make the necessary observations and remarks; another to carry the instruments; another to direct the chain; another to lead it; and another to go forward with a light, as an object, from station to station. During the time of making the survey, be careful in not admitting any iron, steel, or other ferruginous substance, within ten feet of the instrument, for fear of attracting the needle; I have seen the needle affected at almost twice the above distance, by a very massy piece of iron. Also if the glass of the instrument stand in need of cleaning, it must be rubbed as gently as possible, and not with any silken substance, for that will be apt to excite electrical matter, which will prevent the needle from traversing; but if that matter should be excited, it may be very easily discharged, by touching the surface of the glass with the wet finger.

In order for familiarising the young miner with this system of surveying, previous to his practising it in mines, it would be necessary for him to fix up a number of marks on the surface, and afterwards take their bearing and distance from each other, according to the method before directed. But to approach nearer to the form of subterraneous surveying, it would be much better to do it at night, by the assistance of candle-light; many favourable evenings might be found for this mode of practising. Should the current of air be too strong for the naked flame of the candle, lanterns may be used.

To find the magnitude of angles.

(2.) Every circle, ABCD, is supposed to contain 360° (see theorem 2); each semicircle DAB and DCB contains 180°; and each quadrant AB, BC, CD, and DA, contains 90°. Draw the line ab; and if $\angle$ Aab contains 50° $\angle$ Dab must contain 90° — 50° = 40°, and $\angle$ baC must contain 180° — 50° = 130° (see theorem 5). Also if ab makes an angle of 50° with the line AC, and ad an angle of 30° with the same line, the semicircle ADC containing 180°, $\angle$ Aab = 50° + $\angle$ Cad = 30° = 80°, then 180° — 80°, leaves 100° = $\angle$ bad. Or thus, $\angle$ AaD = 90°; then

90° − 50° ∠ A*ab* = 40 ∠ *ba*D; also ∠ D*a*C = 90°; then 90° − 30° ∠ *da*C = 60° ∠ D*ad*; consequently ∠ *ba*D = 40° + ∠ D*ad* = 60° = 100° ∠ *bad*, as before. If *ab* make an angle of 50° with *a*A, and *ac* make another angle of 75° with the same line *a*A, then the ∠ *cab* = 75° − 50° = 25°; and if *ab* make an angle of 50° with *a*A, and *ac* an angle of 25° with the line *ab*, then 50° + 25° = 75° ∠ A*ac*. —(Euc. b. 1, p. 15.)

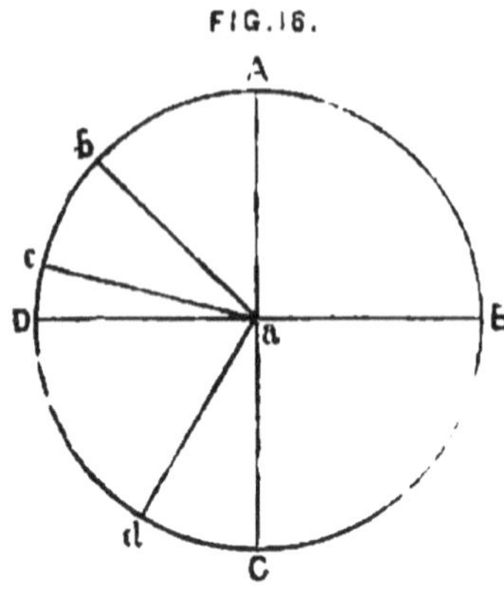

The manner of determining bearings, and also reducing angles into bearings.

(3.) The instrument used in subterraneous surveying is the circumferentor, mentioned as before, whose effect depends on the magnetic needle; and the directions, courses, or bearings, are recorded according to the angles these directions make with the magnetic meridian. (The magnetic meridian is the north and south line, as pointed out by the magnetic needle; see theorem 18.)

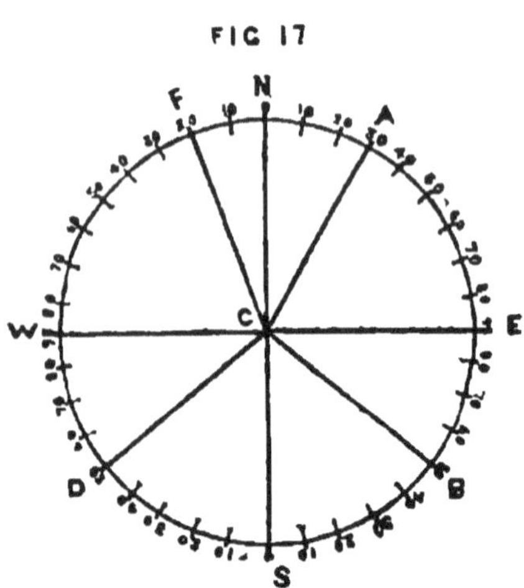

If we pass round from the north N, to the east E, and continue moving from the east to the south S, and from thence to the west W, and lastly from the west to the north N, from whence we first of all set out, we shall have made a circuit NESWN of 360°, which all circles are upposed to contain (see theorem 2); and as there are four cardinal points (see theorem 16) in that circle, north, east, south, and west, dividing it into four equal parts, consequently from north N to east E subtends an angle of 90°; from east E to south S subtends an angle of 90°; from south S to west W subtends an angle of 90°; and from west W to north N subtends an angle of 90°. Now let NE, or the distance between north and east,—also ES, or the distance between east and south,—also SW, or the distance between south and west,—and also WN, or the distance between west and north, be each divided into 90 equal parts or degrees, then a line in direction of CN may be called due north,—and another in direction of CS may be called due south,—another in direction of CE may be called due east, or north 90° east, or south 90° east,—and another in direction of CW may be called due west, or north 90° west, or south 90° west; likewise the line CD passing between S and W, or between south and west, is called south 50° west, being 50° towards the west from south, or to the westward or right-hand (see theorem 17) of the south meridian line. The line CF passing between N and W, or between north and west, is called north 20° west, being 20° towards the west from north; the line CA passing between N and E, or between north and east, is called north 30° east; and the line CB passing between S and E, or between south and east, is called south 50° east; for the bearing of any object from any point or place, taken by the circumferentor, is only the angle that object makes with the magnetic meridian of that point or place from which the bearing is taken: Therefore, if the bearing of B from C is required, it is nothing more than the direction and angle that B makes with the magnetic meridian of C; CS is supposed the magnetic meridian of C, and BCS is the angle the object makes with that meridian.

Let WE, represent a circumferentor, and NS the magnetic needle suspended on the pivot *c* as its centre of suspension and centre of motion; AB are two horizontal arms fixed opposite to each other on the instrument; on the extremity

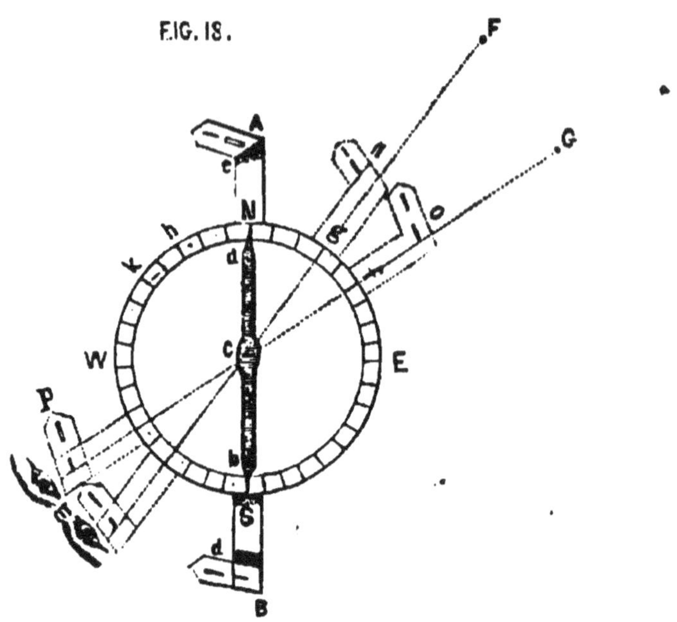

of each arm is the sight *d* and *e* perpendicular thereto, through which is seen the object whose bearing is wanted: The inner part of the circle to which the needle points is divided into degrees, beginning at N, and numbered to 90 each way to W and E; and also beginning at S, and numbered to 90 each way to the same points W and E. The whole of the instrument is fixed on a stand, having a ball and socket to allow of its being kept level and turned freely round. This instrument is manufactured in great perfection by Messrs. Elliott, Brothers, 30, Strand, London.

To find the bearing of the line *c*F, let the centre of the instrument be fixed at *c*; then turning it round so that the eye of the observer may see F through the sights *mn*, the needle always continuing in the same position, or preserving its parallelism, howsoever the instrument and sights are turned, the end N of the needle, which, before the sights were moved, pointed to north N, will, on the sight being moved in direction of *c*F, point to *h* 30°; for *h* will be brought to the situation of N; then the angle N*cg* will be 30°, which is the bearing of F with the north magnetic meridian, and on being found to incline to the right,—therefore, from theorem 17, the bearing of F will be north 30° east, which is usually written N 30° E.

Again, suppose the bearing of G from *c* is wanted, turn the arms and sights *ed* in the situation of *po*, that is, in direction of *c*G, and the angle N*cf* will be 50°,—for the number of 50 at *k* will be opposite the end N of the needle; therefore, from theorem 17, the bearing will be found to be N 50° E.

To find what point of the compass an object bears on, when its direction, with respect to the magnetic meridian, is given.

(4.) Rule.—If the given angle that the object makes with the magnetic meridian is to the right of the north, the object will bear to the east of that meridian; if to the left of that meridian, the object will bear to the west of it. Also, if the given angle that the object makes with the magnetic meridian is to the right of the south, the object will bear to the west of that meridian; if to the left, it will bear to the east of it.—See theorem 17.

Example I.—If I find an object makes an angle to the right of 25° with the magnetic meridian, when I face the north, what is the bearing of that object with that meridian?

From the rule, the object will bear N 25° E.

Example II.—If I find an object makes an angle to the left of 30° with the magnetic meridian, when I face the north, what is the bearing of that object with that meridian?

The object will be N 30° W.

EXAMPLE III.—If I find an object makes an angle to the right of 30° with the magnetic meridian, when I face the south, what is the bearing of that object with that meridian?

The object will bear S 30° W.

EXAMPLE IV.—If I find an object makes an angle to the left of 25° with the magnetic meridian, when I face the south, what is the bearing of that object with that meridian?

The object will bear S 25° E.

EXAMPLE V.—If I find an object makes an angle to the right of 87° with the magnetic meridian, when I face the south, what is the bearing of that object with that meridian?

The object will bear S 87° W.

EXAMPLE VI.—If I find an object makes an angle of 86° to the right with the magnetic meridian, when I face the north, what is the bearing of that object with that meridian?

The object will bear N 86° E.

EXAMPLE VII.—If I find an object makes an angle of 89½° to the right with the magnetic meridian, when I face the north, what is the bearing of that object with that meridian?

The object will bear N 89½° E.

EXAMPLE VIII.—If I find an object makes an angle of 2° to the left with the magnetic meridian, when I face the south, what is the bearing of that object with that meridian?

The object will bear S 2° E.

EXAMPLE IX.—If I find an object makes no angle with the magnetic meridian, when I face the north, what is the bearing of that object with that meridian?

The object will bear due north, or will be in the direction of the magnetic meridian.

EXAMPLE X.—If I find an object makes an angle of

20° to the right of another object which makes an angle of 15° to the right with the magnetic meridian, when I face the north, what is the bearing of the first object with that meridian?

The object will bear (20 + 15 = 35) N 35° E.

EXAMPLE XI.—If I find a line makes an angle of 30° to the right of another line which forms an angle of 60° to the left with the magnetic meridian, when I face the north, what is the bearing of that object with that meridian?

The line will bear (60°−30°=30°) N 30° W.

The reversing of bearings.

(5.) If the bearing of N from S is found to be due north, the bearing of S from N will be due south,—just the reverse of the former; and if the bearing of B from S is found to be N 20° E, the bearing of S from B will be S 20° W,—the reverse; and so of any other.

FIG 19

To reduce bearings into angles.

(6.) Suppose the line CD to bear N 50° E with the magnetic meridian NCS (north being represented by N, and south by S), then CD will make an angle of 50° NCD with the north magnetic meridian CN; and with the south magnetic meridian CS it will make an angle of 180° − 50° NCD = 130° SCD (see theorem 5): And if the line CF bear S 30° E with the magnetic meridian, it will make an angle of 30° SCF with the south magnetic

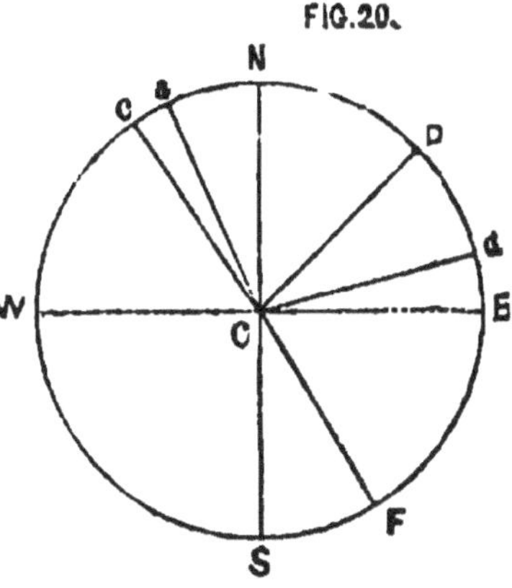

FIG.20.

meridian CS; and with the north magnetic meridian CN it will make an angle of 180° – 30° SCF = 150° NCF: Also if the line CE bear due east with the magnetic meridian, it will make an angle of 90° NCE, or SCE, with that meridian; for east or west always forms an angle of 90° with the meridian (see theorem 16): Or if the line CD bear N 50° E from the point C, and that of C*d* N 80° E from the same point C, then these two bearings being both of the same side of the same meridian, will form an angle with each other of 80° – 50° = 30° ∠ DC*d*: Or if the line CD bear N 50° E from the point C, and that of CFS 30° E from the same point C, then these two bearings, being both of the same side of different meridians, will form an angle with each other of $\overline{180° - 50° + 30°}$ = 100° ∠ DCF: Or if CD bear N 50° E from the point C*a*N 25° W from the same point C, then these two bearings, being of different sides of the same meridian, will form an angle with each other of 50° + 25° = 75° ∠ DC*a*: Or if C*a* bear N 25° W from the point C, and CFS 30° E from the same point C, then these two bearings, being of different sides of different meridians, will form an angle with each other of $\overline{180° - 30°}$ ∠ SCF + 25° NC*a* = 175° ∠ *a*CF, or 30° – 25° = 5°; which difference being taken from 180°, leaves 175° ∠ *a*CF, as before: Or if C*c* bear N 30° W from the point C, and that of CFS 30° E from the same point C, then the one bearing as much to the west side of the north meridian as the other docs to the east side of the south meridian, they will form no angle at all, but a direct line, with each other, or 30° – 30° = 0°; which taken from 180°, leaves 180°, which, as before, shows they form a direct line. (See theorems 4 and 5.)

(7.) If the magnitude of the angles B and C is required, which is formed by the bearing AB taken from A to B, S 50° E; of the bearing BC taken from B to CS 45° W; and of the bearing CD taken from C to D, S 20° E; now to render the two bearings which form the ∠ B to bearings taken from that angular point, the bearing AB, which is the bearing of B from A, must be made the bearing of A from B, by reversing it (Art. 5): Then the ∠ B is the angle formed by the bearings N 50° W, and S 45° W; which bearings, being on the same side of different meridians, will form an angle of $180° - \overline{50° + 45°} = 85°$ ∠ B: And by reversing the bearing BCS 45° W, the ∠ C will be the angle formed by the bearing N 45° ECB, and S 20° ECD; which bearings, being on the same side of different meridians, will form an angle of $180° - \overline{45° + 20°} = 115°$ ∠ C. In determining the magnitude of angles formed by bearings, those bearings which compose the angles must be supposed to be taken from the angular point.

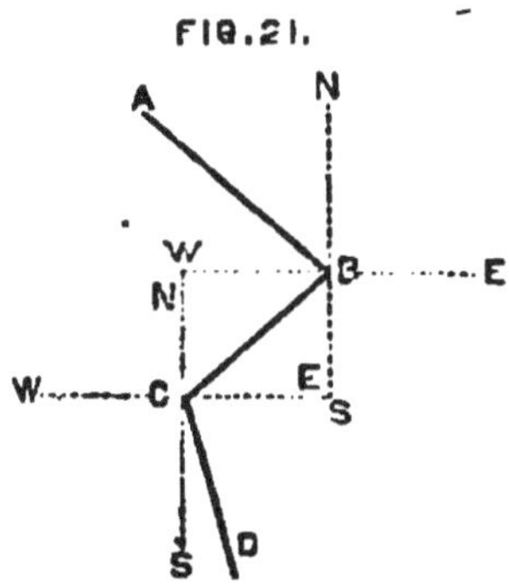

To find the number of degrees contained in the angle that any given bearing makes with the magnetic meridian.

(8.) Rule I.—The number of degrees of the bearing will be the magnitude of the angle that bearing forms with the meridian it is taken from; and the same number of degrees, taken from 180°, leaves the number of degrees contained in the angle the same bearing makes with the contrary meridian.

To find the number of degrees contained in an angle formed by two given bearings taken from the same point, when they are both on the same side of the same meridian.

Rule II.—From the number of degrees contained in the one of the bearings, take the number of degrees contained in the other, and the difference will be the number of degrees contained in the angle formed by the two bearings.

To find the number of degrees contained in an angle formed by two given bearings taken from the same point, when they are both on the same side of different meridians.

RULE III.—Take the sum of the degrees contained in the bearings from 180°, and the remainder will be the number of degrees contained in the angle formed by the two bearings.

To find the number of degrees contained in an angle formed by two given bearings taken from the same point, when they are on different sides of the same meridian.

RULE IV.—The sum of the degrees contained in the bearings is the number of degrees contained in the angle formed by the two bearings.

To find the number of degrees contained in an angle formed by two given bearings taken from the same point, when they are on different sides of different meridians.

RULE V.—Take the difference of the number of degrees contained in the bearing from 180°, and the remainder will be the number of degrees contained in the angle formed by the two bearings.

To find the number of degrees contained in any angle which is formed by given bearings not taken from the angular point.

RULE VI.—Reduce the bearings which compose the angle required into bearings taken from that angular point (which is done by reversing one of them; Art. 5) then by the preceding rules find the number of degrees contained in the angle required.

EXAMPLE I.—In the bearing N 5° W, what is the magnitude of the angle formed with that bearing and the north magnetic meridian?

From rule 1, Art. 8, the bearing will form an angle of 5° with the north magnetic meridian.

EXAMPLE II.—In the bearing S 65° E, what is the

magnitude of the angle with that bearing and the south magnetic meridian?

The bearing will form an angle of 65° with the south magnetic meridian.

EXAMPLE III.—In the bearing S 20° W, what is the magnitude of the angle with that bearing and the north magnetic meridian?

From rule 1, Art. 8, from 180° − 20° = 160°, the magnitude of the angle formed with the north magnetic meridian.

EXAMPLE IV.—In the bearing N 80° E, what is the magnitude of the angle with that bearing and the south magnetic meridian?

From 180° — 80° = 100°, the magnitude of the angle formed with the south magnetic meridian.

EXAMPLE V.—In the bearing due east, what is the magnitude of the angle with that bearing and the south, and also north, magnetic meridian?

The bearing will form an angle of 90° with both the south and north magnetic meridians. (See theorem 16.)

EXAMPLE VI.—In a bearing N 50° E, and another N 80° E, both taken from the same point, what is the magnitude of the angle formed by the two bearings with each other?

From rule 2, Art. 8, the bearings being on the same side of the same meridian 80°—50° = 30°, the magnitude of the angle formed by the two bearings.

EXAMPLE VII.—In a bearing S 60° E, and another S 10° E, both taken from the same point, what is the magnitude of the angle formed by the two bearings with each other?

60°—10° = 50°, the angle that the two bearings will form with each other.

EXAMPLE VIII.—In a bearing S 30° E, and another N 50° E, what is the magnitude of the angle formed by the two bearings with each other, when they are both taken from the same point?

From rule 3, Art. 8, the bearings being on the same side of different meridians, then $180^\circ - \overline{30^\circ + 50^\circ} = 100^\circ$, the magnitude of the angle formed by the two bearings.

EXAMPLE IX. — In a bearing N 80° W, and another S 85° W, what is the magnitude of the angle formed by the two bearings with each other, when they are both taken from the same point?

$180^\circ - \overline{80^\circ + 85^\circ} = 15^\circ$, the magnitude of the angle formed by the two bearings.

EXAMPLE X. — In a bearing N 50° E, and another N 40° W, what is the magnitude of the angle formed by the two bearings with each other, when both taken from the same point?

From rule 4, Art. 8, the bearings being on different sides of the same meridian, then $50^\circ + 40^\circ = 90^\circ$, the magnitude of the angle formed by the two bearings.

EXAMPLE XI.—In a bearing S 10° W, and another S 5° E, both taken from the same point, what is the magnitude of the angle formed by the two bearings with each other?

$10^\circ + 5^\circ = 15^\circ$, the angle formed by the two bearings.

EXAMPLE XII. — In a bearing N 50° E, and another S 30° W, both taken from the same point, what is the magnitude of the angle formed by the two bearings with each other?

From rule 5, Art. 8, the bearing being on different sides of different meridians, then $50^\circ - 30^\circ = 20^\circ$, which taken from $180^\circ = 160^\circ$, the magnitude of the angle formed by the two bearings.

EXAMPLE XIII. — In a bearing S 60° W, and another N 86° E, both taken from the same point, what is the magnitude of the angle they form with each other?

$180^\circ - \overline{86^\circ - 60^\circ} = 154^\circ$, the magnitude of the angle formed by the two bearings.

EXAMPLE XIV. — In a bearing S 80° W, and another N 5° W, both taken from the same point, what is the magnitude of the angle formed with each other?

$180° - \overline{80° - 5°} = 105°$, the magnitude of the angle formed by the two bearings.

Example XV. — In a bearing N 20° W, and another S 20° E, both taken from the same point, what is the magnitude of the angle formed with each other?

$180° - \overline{20° - 20°} = 180°$; therefore the two bearings form no angle, but a direct line, with each other.

Example XVI. — Suppose the bearing ABS 50° E, fig. Art. 7, taken from A to B, and BCS 45° W, taken from B to C, required the magnitude of the angle B formed by those two bearings?

The two bearings which form the angle are not taken from the angular point B, the leg AB being taken from the point A: Therefore, from rule 6, Art. 8, by reversing the bearing ABS 50° E, to BAN 50° W, the angle B will then be formed of two bearings taken from the same angular point (viz.) BAN 50° W, and BCS 45° W; which bearing on the same side of different meridians, from rule 3, Art. 8, will be $180° - \overline{50° + 45°} = 85°$, the magnitude of the required angle.

Example XVII. — In the bearing BCS 45° W, fig. Art. 7, taken from B to C, and CDS 20° E taken from C to D, required the magnitude of the angle C formed thereby?

The two bearings which form the angle are not taken from the angular point C; Therefore, from rule 6, by reversing BCS 45° W to N 45° E, the angle C will then be composed of two bearings taken from that angular point, N 45° E, and S 20° E; which bearing on the same side of different meridians, from rule 3, Art. 8, will be $180° - \overline{45° + 20°} = 115°$, the magnitude of the required angle.

The reducing of bearings and distances to their northing or southing, and easting or westing, from the point of departure.

(9.) Suppose it is required to know how far B is northward of A. If SN represent the meridian, Aa will be the northing of B from A, and aB, or Ab, will be its easting, or what B is eastward of the meridian of the point A.

FIG. 22

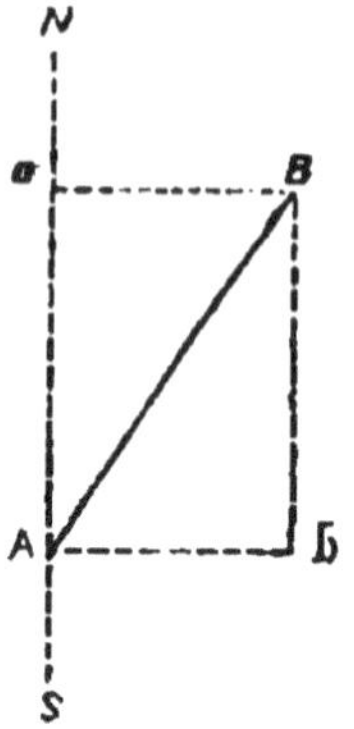

If we pass from any point A to B (fig. 23), and from B to C, and from C to D, and from thence return to the point A again, our route will have as much southing as northing, and easting as westing; that is, the southing will be equal to the northing, and the easting will be equal to the westing. Let SN represent the meridian of A, the point of departure; the northing of AB from that point will be represented by Aa, the northing of ABC by Ab, that of ABCD by Ac; and as the northing of D from A is equal to Ac, the southing of A from D will be equal to $De = cA$; therefore the southing will be equal to the northing of ABCDA: Also let the westing of AB from the meridian of A be represented by aB, that of ABC by bC, that of ABCD by cD; then as the westing of D from the meridian of A is equal to cD, the easting of A from D will be equal to $eA = cD$: Therefore the easting will be equal to the westing of ABCDA.

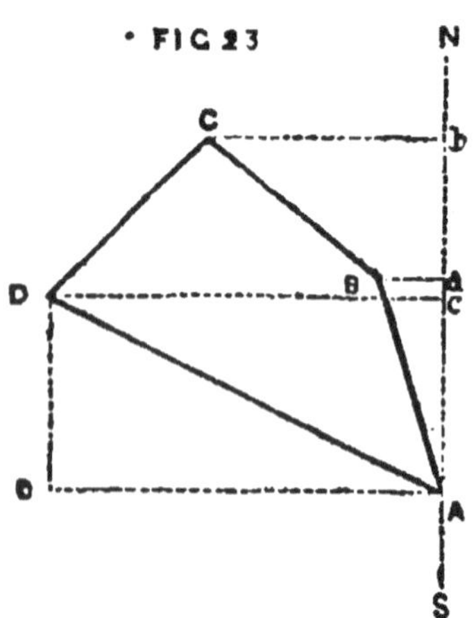

FIG. 23

To find the northing or southing, and also the easting or westing, of any given bearing and distance.

(10.) RULE.—Look in the traverse tables, under the degree answering to that of the bearing: and to the right, opposite the length of that bearing (in the column of distances), will be found the quantity of northing or southing, according as the bearing is north or south,—and also the easting or westing, according as the bearing is east or west.

Note.—When the given distance consists of chains and links, and the angles of degrees and minutes, the method of finding the northing or southing, and the easting or westing, is given in Art. 59.

EXAMPLE I.—What is the northing and easting of the bearing and distance N 40° E 10 chains?

Look in the traverse tables, under 40°, and opposite 10 chains, in the column of lengths, is found 7 chains 60 links of northing, and 6 chains 43 links of easting (the links are usually written after the chains as decimals, see next Example).

EXAMPLE II.—What is the northing and westing of N 10° W 6·50 chains from the point of commencement, or point of departure?

Its northing is 6·40 chains, and its westing is 1·13 chains.

EXAMPLE III.—What is the southing and westing of S 79° W 7·30 chains from the point of departure?

Its southing is 1·39 chains, and its westing 7·16 chains.

EXAMPLE IV.—What is the southing and westing of S 80° W 6 chains?

Its southing is 94 links, and its westing 5·36 chains.

EXAMPLE V.—What is the northing and westing of N 40° W 8·50 chains?

Its northing is 6·50 chains, and its westing is 5·46 chains.

EXAMPLE VI.—What is the southing and easting of S 5° E 6·52 chains?

Its southing is 6·49 chains, and its easting is 56 links.

EXAMPLE VII.—In the following successive bearings and distances taken from the pit A to the pit B, N 50° W 10 chains, N 20° E 5 chains, and S 40° W 7 chains, I wish to know what denomination of bearing the pit B will have from A; also the length of each?

PREPARATORY TABLE.

BEARINGS.	NORTHING	SOUTHING.	EASTING.	WESTING.
	Chains.	Chains.	Chains.	Chains.
N 50° W 10 chains	6·43	...	...	7·66
N 20° E 5 ,,	4·70	...	1·71	...
S 40° W 7 ,,	...	5·36	...	4·50
	11·13			12·16
	5·36			1·71
	5·77			10·45

Now, as the northing is greater than the southing by 5·77 chains, and the westing greater than the easting by 10·45 chains, the pit will have 5·77 chains of northing, and 10·45 chains of westing from A.

Example VIII.—What is the northing and westing of the pit D from C, under the following bearings, N 20° E 10 chains, and S 60° W 6 chains?

BEARINGS.	NORTHING.	SOUTHING.	EASTING.	WESTING.
	Chains.	Chains.	Chains.	Chains.
N 20° E 10 chains	9·40	...	3·42	...
S 60° W 6 ,,	...	3·00	...	5·20
	9·40			3·42
	3·00			1·78
	6·40			

The southing being taken from the northing, and the easting from the westing, the pit D will have 6·40 chains of northing, and 1·78 chains of westing, from the pit C.

Example IX.—What is the northing and easting of B from A under the following successive bearings, S 20° W 10 chains, N 5 chains, N 30° E 20 chains, and N 5° E 8 chains?

The pit B will have 20·89 chains of northing, and 7·28 chains of easting, from A.

Surveying and recording bearings.

(11.) Suppose the bearing of ABC to be required. Set the circumferentor on A (the north being represented by N and the south by S): then turning that part of the instrument having the *fleur-de-lis* from you, or towards B, turn the instrument until the object B is seen through, and cut by, the hair in the sights; and the angle NAB being the angle that the sights and line AB make with the magnetic meridian NS, will be the bearing of B from A,—suppose 30°; which also being to the right side of the north meridian, will be N 30° E (see theorem 17): Then bring the instrument forward to B, and fix it there, directing the same sight at B towards C as was directed at A towards B; then observe the angle that BC makes with the magnetic meridian,—which suppose 25° NBC; and being to the left of the meridian, will be N. 25° W. In order to prove the work, and try the accuracy of the instrument when it is standing at B, apply the eye to that sight which was next B when it stood at A; then take the bearing of A from B, which, if found to be the reverse of B from A, shows the work so far is true. The bearing of B being taken, in like manner, from C, will prove the truth of the survey. Observe always to take the degrees of each bearing by the same end of the needle.

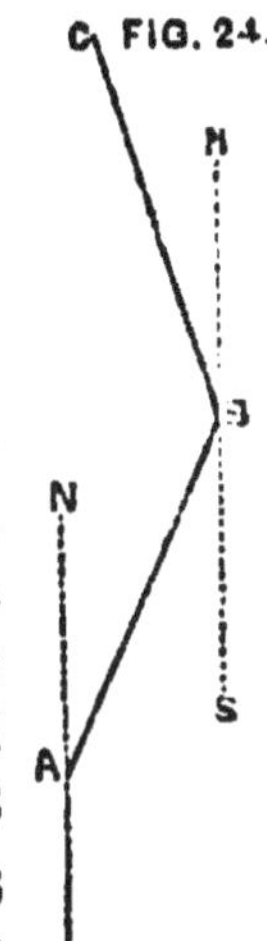

(12.) Suppose the bearing of B from A, C from B, and D from C, to be required: Fix the instrument at A, with the *fleur-de-lis* towards B (the north being represented by N and the south by S); then take the bearing of B, as before-described,—which suppose to make an angle of 30°, NAB to the right with the magnetic meridian, or N 30° E; remove the instrument to B, and take the bearing of C,—which suppose equal to 30° NBC to the left, or N 30° W; then remove the instrument to C, and take the bearing of D,—which suppose equal to 65° SCD to the left, or S 65° E: See below in the survey-book.

FIG. 25.

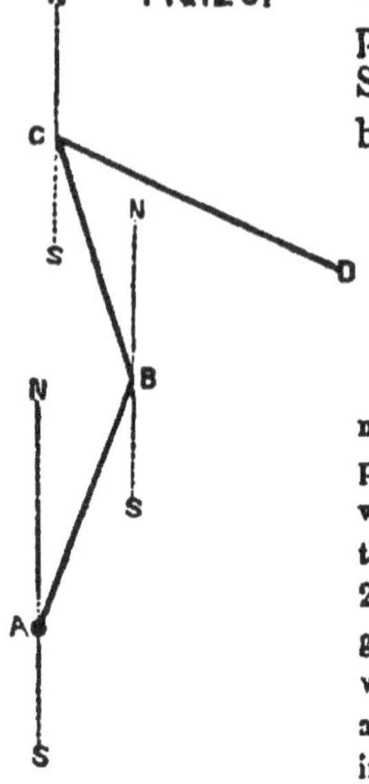

From A to B, N. 30° E.
B to C, N. 30° W.
C to D, S. 65° E.

Note.—This survey may be proved in the same manner as the preceding; and the methods of plotting and protracting subterraneous surveys, with the descriptions of the instruments used for the purpose, are given in Arts. 23, 24, 25 and 26, to which the student is referred, as it would greatly facilitate in these studies to lay down his work on paper, as soon after he has finished it as a proper opportunity presents itself; all the instruments required, in the first instance, being a scale of equal parts, a protractor, and a T-square.

(13.) Suppose the subterraneous working ABCDA, to be surveyed, beginning at the pit A: Fix the instrument at the centre of the pit A; then let a person hold a lighted candle at B, being the utmost distance it can be seen through the sights of the instrument, the bearing of which being taken from A, suppose due south, or in the direction of the magnetic meridian of A,—and its distance from A suppose 6·57 chains; which place in the survey-book, as below:

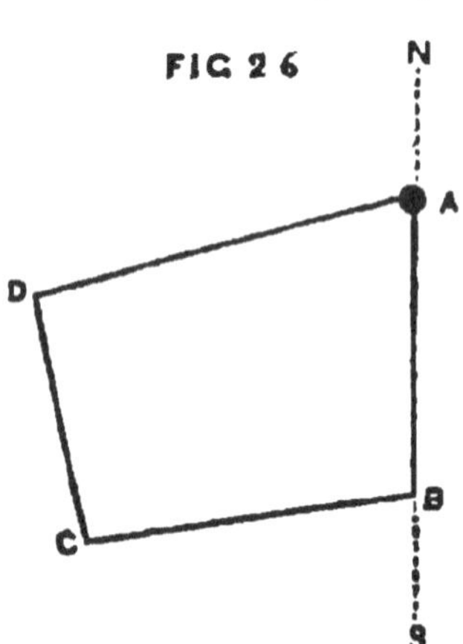

FIG 26

Remove the instrument to B, where the candle stood, and direct the person to place the lighted candle at C; then take its bearing from B, which suppose to make an angle CBS of 80° with the magnetic meridian, or to bear S 80° W; and its distance being found 7·10 chains, remove the instrument to C, the lighted candle being removed to D; then take its bearing and distance as before, which suppose N 10° W 5 chains; remove the instrument to D, and direct the lighted candle to be placed at the centre of the pit A, where the survey commenced; then take its bearing from D, which suppose N 70° E 8·35 chains,—and the survey will be finished.

A survey of a subterraneous working, commencing at the centre of the pit A.

AB, S. . . .	6·57 chains.
BC, S. 80° W. . . .	7·10 ,,
CD, N. 10° W. . . .	5·00 ,,
DA, N. 70° E. . . .	8·35 ,,

This survey, which is composed of four sides, may be proved by adding together the degrees contained in the interior angles, which, if they amount to 360, the work will be right.—(See theorem 11).

The proof.—The magnitude of the angle DAB is 70° (see rule 1, art. 8) in the reducing of bearings into angles; angle ABC is 180° − 80° ∠ CBS = 100°; angle BCD is 80° + 10° = 90° (see rule 4, art. 8); and angle CDA is $180° - \overline{70° + 10°} = 100°$ (see rule 3, art. 8).—

$$
\begin{aligned}
\text{Then } \angle \text{DAB} &= 70° \\
\angle \text{ABC} &= 100° \\
\angle \text{BCD} &= 90° \\
\angle \text{CDA} &= 100° \\
\hline
&\ 360°
\end{aligned}
$$

Also the proof may be made by finding the northing, southing, easting, and westing of all bearings and distances. If the southings are equal to the northings, and the westings equal to the eastings, then the work will be right.—(See Art. 9.)

	Northing.	Southing.	Easting.	Westing.
C. L.	Chains.	Chains.	Chains.	Chains.
Thus, S. . . . 6 57	...	6·57	...	...
S. 80° W. 7 10	...	1·23	...	6·98
N. 10° W. 5 0	4·93	...	...	0·87
N. 70° E. 8 35	2·87	...	7·85	...
	7·80	7·80	7·85	7·85

Therefore the northings and southings being equal, as also the eastings and westings equal, the work is right.

(14.) Suppose the bearing and distance of B from A, C from B, D from C, F from D, G from B, H from G, and I from H, are required. Fix the instrument at A, and take the bearing (as before described) of B from it, which suppose to be N 30° W, and the distance 5·50 chains; set it down in the following survey-book; also make a mark * with chalk at B, which must likewise be noted down, to return to. In order to take the bearings of C, D, and F, remove the instrument to B, and take the bearing of C from it, which suppose N 45° E, and distance 7 chains; the bearing of D from C suppose N 50° W, and the distance 5 chains; the bearing of F from D suppose N 85° E, and distance 7 chains: Then bring the instrument from D to the chalk mark at B, and take the bearing of G from B, which suppose S 65° W, and distance 6·50 chains; the bearing of H from G suppose N 10° W, and distance 6 chains; and lastly, the bearing of I from H suppose N. 60° E., and distance 4 chains.—See them properly arranged in the following survey-book (p. 29).

Suppose the bearings and distances of B from the pit A, C from B, D from C, F from D, G from F, H from G, P from H, and A from P,—also I from O, K from I, L from I, and M from L, are required, together with any remarkable circumstance that may be met with in the survey: Fix the instrument at A, directing a person to go with a lighted candle to B, and take the bearing and

SURVEY-BOOK.

Commencing at A.	Chains.	
N. 30° W. . .	5·50	to B.
At B is a chalk mark * to return to.		
N. 45° E. . .	7·00	to C.
N. 50° W. . .	5·00	to D.
N. 85° E. . .	7·00	to F.
Returns to the chalk mark * at B, and proceeds to G, &c.		
S. 65° W. . .	6·50	to G.
N. 10° W. . .	6·00	to H.
N. 60° E. . .	4·00	to I.

FIG. 27.

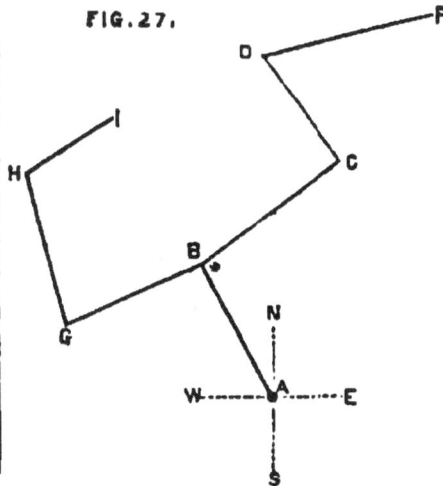

distance of B from it, which suppose S 36° E 7 chains; which insert in the survey-book. Also at *a*, 3 chains from

FIG. 28.

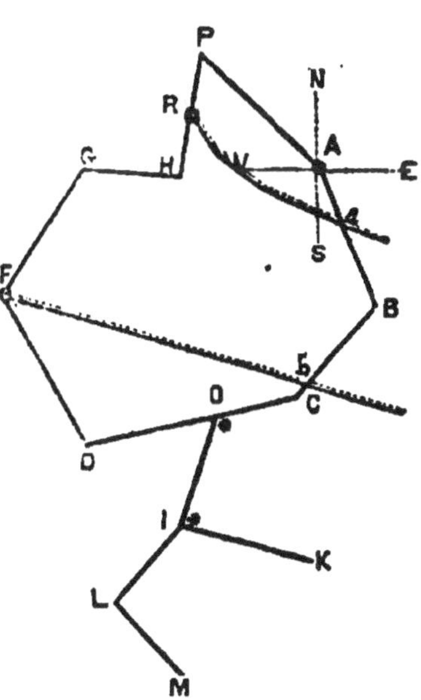

A towards B, is the water-course from the pit R: Bring the instrument to B, and take the bearing and distance of the lighted candle at C, which suppose S 42° W 4 chain. At *b*, 3 chains from B towards C, is the water-course from the pit F: Remove the instrument to C, and take the bearing and distance of the light at D, which suppose S. 75°. W. 10 chains. At 4 chains from C towards D is a chalk mark * at O, to return to. Take the bearing and distance of F from D, which suppose N 42° W 7·50 chains, to a pit; which note also down in the survey-book: Take the bearing and distance of G from F, which suppose N

42° E 5 chains: Take the bearing and distance of H from G, which suppose E 4 chains: Take the bearing and distance of P from H, which suppose N 9° E 4 chains. At 2 chains from H, towards P, is the pit R; which note down in the survey-book. Take the bearing and distance of A from P, which suppose S. 69° E. 5·56 chains. Now return to the * mark at O, and fix the instrument there: Take the bearing and distance of the candle placed at I, which suppose S 10° W, 5 chains, and a chalk mark * to return to: Take the bearing and distance of K from I, which suppose S 80° E 6 chains. Return to the mark at I, and fix the instrument, and take the bearing and distance of L from it, which suppose S 40° W 4 chains: Take the bearing and distance of M from L, which suppose S 45° E 4 chains,—and the survey will be finished.

See the bearings and distances arranged in form of a Survey-book.

		Chains.	
From A to B	S. 36° E.	7·00	
	At	3·00	from A is *a*, the water-course from the pit R.
B to C	S. 42° W.	4·00	
	At	3·00	from B is *b*, the water-course from the pit F.
C to D	S. 75° W.	10·00	
	At	4·00	from C is O, a chalk mark * to return to.
D to F	N. 42° W.	7·50	and pit F.
F to G	N. 42° E.	5·00	
G to H	E. . .	4·00	
H to P	N. 9° E.	4·00	
	At	2·00	from H is the pit R.
P to A	S. 69° E.	5·56	
			Return to the chalk mark * at O.
O to I	S. 10° W.	5·00	
			At I is a chalk mark * to return to.
I to K	S. 80° E.	6·00	
			Returned to the chalk mark * at I.
I to L	S. 40° W.	4·00	
L to M	S. 45° E.	4·00	

(15.) Suppose a survey of the subterraneous working ABCDFGHPA see last fig., is required, to commence at the pit A: Proceed as in the last, recording each bearing and its respective distance in the following manner, as in the survey-book:—

THE SURVEY COMMENCING AT THE PIT A.

		Chains.
From A to B	S. 36° E.	7·00
B to C	S. 42° W.	4·00
C to D	S. 75° W.	10·00
D to F	N. 42° W.	7·50
F to G	N. 42° E.	5·00
G to H	E. . . .	·4·00
H to P	N. 9° E.	4·00
P to A	S. 69° E.	5·56

To prove this survey by theorem 12: The number of sides which the survey contains is 8; then the amount of all the angles contained in the figure is equal to $\overline{8 \times 2 \times 90^\circ} - \overline{4 \times 90^\circ} = 1080^\circ$. Now from the rules of reducing bearings into angles (art. 8), ∠ B = 102°, ∠ C = 147°, ∠ D 117°, ∠ F = 96°, ∠ G 132°, ∠ H = 261°, ∠ P = 78°, and ∠ A = 147°; whose sum is equal to 1080°, as before: Therefore the survey is right.

Also the same may be proved by taking the northing, southing, easting, and westing of the bearings (art. 9); which, if the southings are found equal to the northings, and the westings equal to the eastings, the survey will be right.

Bearing and Distance.	Northing.	Southing.	Easting.	Westing.
Chains.	Chains.	Chains.	Chains.	Chains.
S. 36° E. 7·00	...	5·66	4·21	...
S. 42° W. 4·00	...	2·97	...	2·68
S. 75° W. 10·00	...	2·59	...	9·66
N. 42° W. 7·50	5·55	...	...	4·97
N. 42° E. 5·00	3·72	...	3·35	...
E.	...	...	4·00	...
N. 9° E. 4·00	3·95	...	0·63	...
S. 69° E. 5·56	...	2·00	5·21	...
	13·22	13·22	17·31	17·31

The northings and southings being equal, and the eastings and westings being also equal, the survey must be right.

(16.) Suppose the subterraneous bearings and distances of the workings ABCDF, are required, commencing at the pit A; and likewise the bearing and distance on the surface of a sinking pit G from the pit A: Fix the instrument at the centre of the pit A in the mine, and take the bearing and distance of B from it; also, of C from B, D from C, and F from D; then the survey under-ground will be completed. Ascend to the surface, and fix the instrument at the mouth of the pit A; and having previously placed a mark at the pit G, take its bearing and distance from A, which insert in the survey-book. (Observe to take the bearing of G from the same point of the pit A, on the surface, that the subterraneous survey commenced under the surface,—so that the proper situation of F, or any other part of the subterraneous working, may be shown with

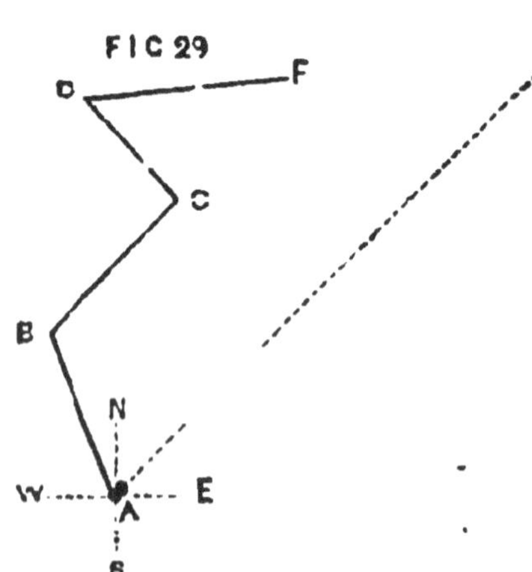

respect to the two pits.) If the bearing and distance of the pit G from that of A cannot be got at once, by the interposition of any building or other obstruction, it must be taken at two or three, or more, different bearings.

THE SURVEY COMMENCING AT THE PIT A.

		Chains.
From A to B	N. 30° W.	5·50
B to C	N. 45° E.	7·00
C to D	N. 50° W.	5·00
D to F	N. 85° E.	7·00
	The bearing and distance of the sinking pit G from the pit A, taken on the surface.	
A to G	N. 45° E.	18·00

(17.) Suppose the subterraneous workings CDFGHI KBLMOP are required to be surveyed, beginning at the pit A: Fix the instrument at A, and take the bearing and distance of the headways AC (as before shown), which suppose S 10° E 3·10 chains. At 80 links is a bord 1 to the right and left, holed into the headways each way; at 1·60 chains is a bord 2 to the right and left, holed each way; at 2·40 chains is a bord 3 to the right, 1 chain to the face, and to the left holed into the headways. Take the bearing and distance of A*a*, which suppose S 80° W 1·60 chains: At 1·30 chains is a headways R to the right and left, and a mark * to return to. Take the bearing and distance of *a*G, which suppose S. 70° W. 1·80 chains: At 80 links is a headways *b* to the right, and a mark * to return to; and at 1·20 chains is a headways V to the left, and a mark + to return to. Take the bearing and distance of the headways RD (by fixing the instrument at the mark at R), which suppose S 8° W 2·50 chains: At 70 links is a bord 4 to the right and left, and holed each way; at 1·50 chains is a bord 5 to the right and left, and holed each way. Take the bearing and distance of the headways VF (by fixing the instrument at the mark at V), which suppose

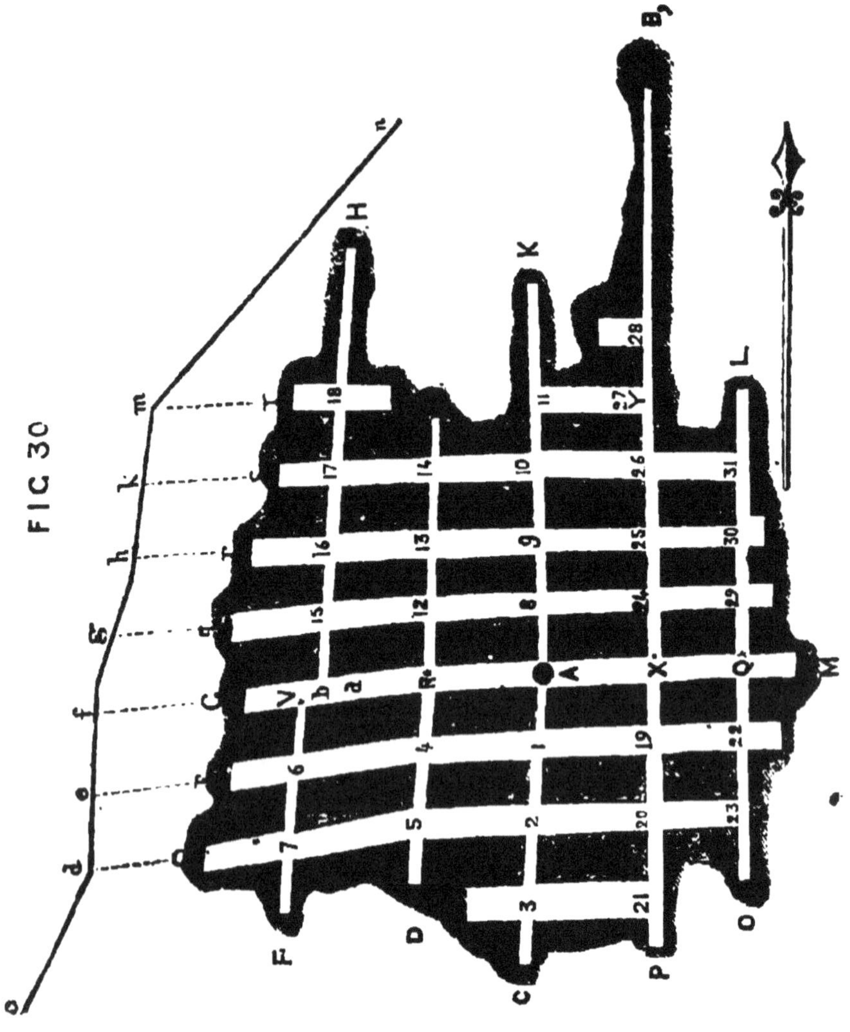

FIG 30

S 10° W 2·40 chains: At 80 links is a bord 6 to the right 1·30 chains to the face, and to the left holed into the headways; at 1·60 chains is a bord 7 to the right 1 chain to the face, and to the left holed into the headways. Take the bearing and distance of the headways AK (by fixing the instrument at A), which suppose N 10° W 4·20 chains: At 80 links is a bord 8 to the right and left, and holed into the headways each way; at 1·70 chains is a bord 9 to the right and left, and holed into the headways each way; at 2·50 chains is a bord 10 to the right and left, and holed into the headways each way; at 3·30 chains is a bord 11 to the right, and holed into the headways,—and none to the left. Take the bearing and distance of the headways RI (by fixing the instrument at the mark at R), which suppose N 2° W 3 chains: At 80 links is a bord 12 to the right and left, and holed into the headways each way; at 1·60 chains is a bord 13 to the right and left, and holed into the headways each way; at 2·40 chains is a bord 14 to the right and left, and holed into the headways each way. Take the bearing and distance of the headways *b*H (by fixing the instrument at the mark at *b*), which suppose N 1° W 5 chains: At 80 links is a bord 15 holed into the headways to the right, and to the left 90 links to the face; at 1·70 chains is a bord 16 holed into the headways to the right, and to the left 60 links to the face; at 2·55 chains is a bord 17 holed into the headways to the right, and to the left 60 links to the face; at 3·40 chains is a bord 18 to the right 50 links to the face, and to the left 55 links to the face. Take the bearing and distance of AM (by fixing the instrument at A), which suppose N 85° E 2·80 chains: At 1·30 chains is a headways X to the right and to the left, and a mark + to return to; at 2·50 chains is a headways Q to the right and to the left, and a mark * to return to. Take the bearing and distance of the headways XP (by fixing the instrument at the mark X), which suppose S 5° E 3·10 chains: At 75 links is a bord 19 to the right and left, and holed into the headways each way; at 1·60 chains is a bord 20 to the right and left, and holed into the headways each way; at 2·40 chains is a bord 21 to the right, and holed into the headways,—and none to the left. Take the bearing and distance of the headways

QO (by fixing the instrument at the mark Q), which suppose S 4° E 2 30 chains: At 80 links is a bord 22 to the right, and holed into the headways, and to the left 40 links to the face; at 1·60 chains is a bord 23 to the right, holed into the headways,—and none to the left. Take the bearing and distance of the headways XYZB (by fixing the instrument at the mark X), which suppose from X to Y N 2° W 2 80 chains: At 90 links is a bord 24 to the right and left, and holed into the headways each way; at 1·70 chains is a bord 25 to the right and left, and holed into the headways each way; at 2·60 chains is a bord 26 to the right and left, and holed into the headways each. Take the bearing and distance of YZ (by fixing the instrument at Y), which suppose N 5° W 2 chains: At 60 links is a bord 27 to the left, holed into the headways,—and none to the right; at 1·50 chains is a bord 28 to the left 30 links, and none to the right. Take the bearing and distance of ZB (by fixing the instrument at Z), which suppose N 3° W 2·30 chains, to a pit B. Lastly, take the bearing and distance of the headways QL (by fixing the instrument at the mark Q), which suppose N 2° W 3·60 chains: At 80 links is a bord 29 to the left, holed into the headways, and to the right 30 links to the face; at 1 link is a bord 30 to the left, holed into the headways, and to the right 20 links to the face; at 2·60 chains is a bord 31 to the left, holed into the headways, and none to the right.—See the survey-book, where the whole is recorded:—

SURVEY-BOOK.

A SURVEY OF A PIT'S WORKINGS, COMMENCING AT THE PIT A.

Bearings.	Remarks to Left.	Dist.	Remarks to Right.	
		Chains.		
S. 10° E.	. . . . . .	3·10	. . . . .	AC
	Bord holed . .	0·80	Bord holed	
	Bord holed . . .	1·60	Bord holed	
	Bord holed . .	2·40	Bord 1 chain from the headways	
S. 80° W.	. . . . . .	1·60	. . . . .	A*a*
	Headways . .	1·30	Headways	
	And a chalk mark * at R to return to	.		
S. 70° W.	. . . . . .	1·80	. . . . .	*a*G
		0·80	A headways *b*, and a chalk mark * to return to	
	Headways V, and a chalk mark * to return to	1·20		
S. 8° W.	. . . . . .	2·50	. . . . .	RD
	Bord holed . .	0·70	Bord holed	
	Bord holed . . .	0·50	Bord holed	
	Returned to the mark * at V			
S. 10° W.	. . . . . .	2·40	. . . . .	VF
	Bord holed . . .	0·80	Bord 1·30 chain from the headways	
	Bord holed . .	1·60	Bord 1 chain from the headways	
	Returned to the pit A			
N. 10° W.	. . . . . .	4·20	. . . . .	AK
	Bord holed . .	0·80	Bord holed	
	Bord holed . . .	1·70	Bord holed	
	Bord holed . .	2·50	Bord holed	
	None	3·30	Bord holed	
	Returned to the mark * at R			
N. 2° W.	. . . . . .	3·00	. . . . .	RI
	Bord holed . . .	0·80	Bord holed	
	Bord holed . .	1·60	Bord holed	
	Bord holed . . .	2·40	Bord holed	
	Returned to the mark * at *b*			
N. 1° W.	. . . . . .	5·00	. . . . .	*b*H
	Bord 90 links from the headways	0·80	Bord holed	
	Bord 60 links from the headways	1·70	Bord holed	

Bearings.	Remarks to Left.	Dist.	Remarks to Right.	
		Chains.		
	Bord 60 links from the headways	2·55	Bord holed	
	Bord 55 links from the headways	3·40	Bord 50 links from the headways	
	Returned to the pit A			
N. 85° E.	. . . . . .	2·80	. . . . .	AM
	Headways . .	1·30	Headways	
	And a chalk mark * at X to return to			
	Headways . .	2·50	Headways	
	And a chalk mark * at Q to return to			
	Returned to the mark * at X			
S. 5° E.	. . . . . .	3·10	. . . . .	XP
	Bord holed . .	0·75	Bord holed	
	Bord holed . . .	1·60	Bord holed	
	None . . .	2·40	Bord holed	
	Returned to the mark * at Q			
S. 4° E.	. . . . . .	2·30	. . . . .	QO
	Bord 40 links from the headways	0·80	Bord holed	
	None . . .	1·60	Bord holed	
	Returned to the mark * at X			
N. 2° W.	. . . . . .	2·80	. . . . .	XY
	Bord holed . .	0·90	Bord holed	
	Bord holed . . .	1·70	Bord holed	
	Bord holed . .	2·60	Bord holed	
N. 5° W.	. . . . . .	2·00	. . . . .	YZ
	Bord holed . .	0·60	None	
	Bord 30 links from the headways	1·50	None	
N. 3° W.	. . . . . .	2·30	To pit B . .	ZB
	Returned to the mark * at Q			
N. 2° W.	. . . . . .	3·60	. . . . .	QL
	Bord holed . .	0·80	Bord 30 links from the headways	
	Bord holed . . .	1·70	Bord 20 links from the headways	
	Bord holed . .	2·60	None	

Note.—When marks are made to be returned to in the survey, observe that they are returned to, otherwise the survey will be defective; and when the new method of taking the angles, given in Arts. 20 and 21, is adopted, the angles, thus taken, must be inserted instead of the bearings, the column being headed "angles" instead of "bearings."

The Back-Sight.

(18.) Suppose the bearing and distance of B from the pit A is required: Fix the instrument at B, instead of A (keeping the same sight foremost, and pointing towards *b*, when it is placed in the situation of B, as if it had been placed in the situation of A, for the purpose of taking the bearing of B); then apply the eye at the sight furthest distant from A, turning the same until the light at the pit A is cut by the perpendicular hair in the other; observe then the bearing of A from B, which, suppose S 30° E, on being reversed (see p. 17), will become N 30° W, for the bearing of B from A,—the distance, being measured, is found to be 3 chains; making the bearing and distance of B from A N 30° W 3 chains.

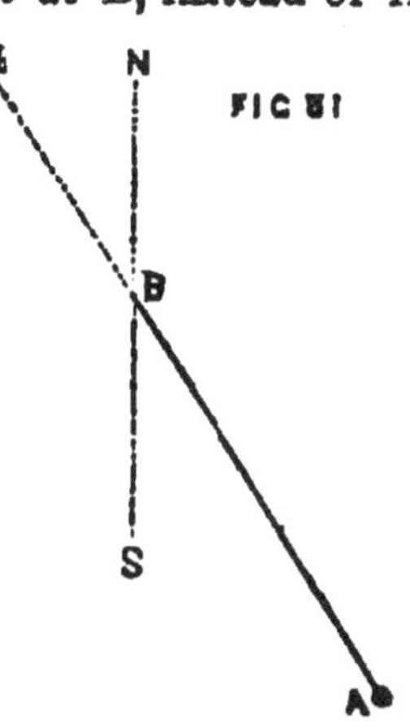

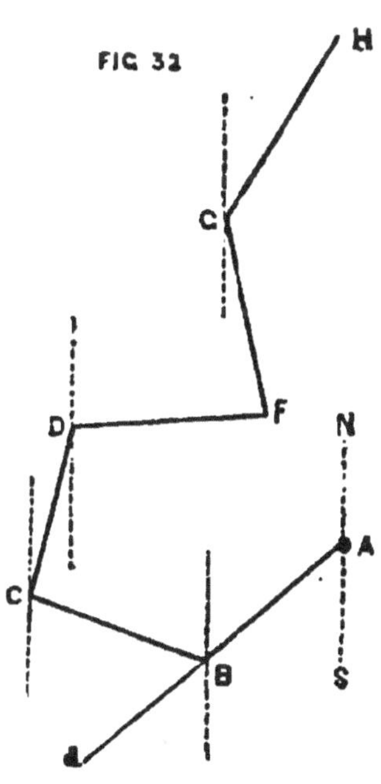

Bearings taken in this way are taken in a direction contrary to the order of the survey, and the eye is applied at the contrary sight to that which it would be applied when direct bearings are taken.

(19.) Suppose the bearing of ABCDFG and H is required, making use of the back-sight throughout the survey: Fix the instrument at B, instead of A, directing that sight towards A which, in the situation of A, would have been hindmost, in the manner before directed; then the bearing A from B being found to be N 45° E, on being reversed makes S 45° W, the bearing of B from A, which enter into the survey-book. The instrument standing at B, turn that sight towards C which pointed to *a*, and take the bearing of C from B, which being found N 75° W, enter the same into the survey-book, without reversing, as it is not a backsight. Remove the instrument from B to D, and direct the sight back to C from D, in the same manner as from B to A: The bearing then of C from D being found S 10° W, which, being reversed, will be N 10° E, the bearing of D from C,—which enter into the survey-book. Then take the bearing of F from D, which being found N 80° E, enter the same into the survey-book, without reversing. Lastly, remove the instrument to G; then take the back-sight from G to F, which being found S 15° E, on being reversed will be N 15° W, the bearing of G from F,—which being entered into the survey-book, then take the bearing of H from G; which suppose N 30° E,—which enter also into the survey-book, without reversing, and the survey is finished.—By this mode of taking bearings, the instrument is only removed half the number of times it would otherwise be, were the back-sights not taken.

SURVEY-BOOK.

The bearing of B from A, S. 45° W.
,, C from B, N. 75° W.
,, D from C, N. 10° E.
,, F from D, N. 80° E.
,, G from F, N. 15° W.
,, H from G, N. 30° E.

PART II.

ON SURVEYING SUBTERRANEOUS EXCAVATIONS WITHOUT THE GENERAL USE OF THE NEEDLE.

It has long been found that the conducting of subterraneous surveys requires strict attention in guarding against the presence of ferruginous substances, which exist in almost all mines, and which, it is well known, affect the magnetic needle, so as to cause it to give erroneous indications. On this account Mr. Fenwick was induced, as long ago as 1822 (when the second edition of his work was published), to suggest to the surveyor of mines the following new method, in which the needle has no control except in the first departure.

*The use of the instrument.**

Suppose the subterraneous excavation ABCDEF to be surveyed beginning at the pit A, and terminating at the pit F.

(20.) Place the instrument at B, and turning it until the vanes at zero cut the lighted candle at the centre of the pit A, which suppose N 65° E; and suppose AB to be 3 chains, the fixed sight at 0° still remaining as before; screw the instrument fast, and turn the moveable sights so as to cut a candle placed at C, taking care that the instru-

* The improved Circumferentor, by Elliott Brothers, 30, Strand, London, which is still much used, especially in secondary mining surveys; but the modern improved theodolite is much to be preferred. See *Heather's Treatise on Mathematical Instruments*, *Weale's Series*.

ment has remained immovable. If so, read off the angle, which the index makes with the moveable circle, which sup-

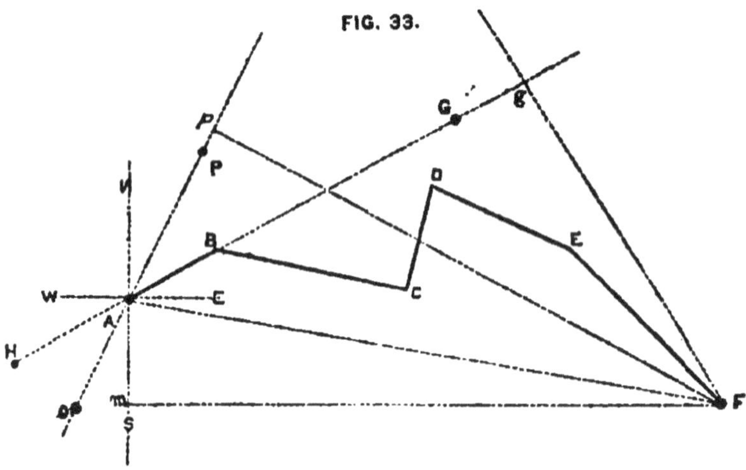

FIG. 33.

pose 120°; then the angle ABC is 120°, that is, the excavations BC and AB make an angle of 120°. Removing the instrument to C, turn the sights and index so as to cut the candle at B, keeping the instrument immovable; then turn the sights to the candle at D, reading off the angle BCD, which suppose 80°; and measure CB, which call 5 chains. Remove the instrument next to D, measuring the distance CD, which call 3 chains; and turn the sights and index to the candle at C, the instrument, as before, being kept immovable; turn the sights to the candle E, observe the angle CDE, which suppose 70°, and let the distance DE be 4 chains. Remove the instrument to E, and turn the sight and index to the candle at D, keeping the instrument immovable; turn the sight to the candle at E, and observing the angle DEF = 160°; lastly, measure the distance EF, which call 6 chains, and the survey is completed.

The following method of conducting subterraneous survey entirely without the use of the magnetic needle, was suggested by Mr. T. Baker (who has now made the present additions and improvements to the new edition of Mr. Fenwick's "Subterraneous Surveying"), at least 35 years ago; but it was ridiculed by the then colliery surveyors; yet is now recommended and adopted by several scientific mining surveyors; among whom I may name Mr. H. Mackworth; who has given more elaborate details for conducting these surveys than those in the preceding article: Mr. M.'s improvements on Mr. Baker's suggestion are given in the following article.

(21.) To commence a survey without the magnetic needle, where there is only one shaft to the mine, the following plan should be adopted. Two thin copper wires, carrying heavy weights, must be suspended from a strong straight edge, at the surface of the shaft, and as near the edges of the shaft as not to touch them, the weights reaching nearly to the bottom of the shaft; while the weights must be immersed in buckets of water, or what would be still better, in vessels filled with mercury, to diminish oscillation, which will still continue, if the shaft is deep; but in the latter case, for only a very short time. The observer standing behind the wires must next send a candle along the heading, as far as it can be seen, and have it fixed in a line with the wires. He should repeat the operation in the opposite direction, by placing a candle against one of the wires, that the whole may be checked by seeing that the three candles are exactly in a line. This line being the basis of the whole underground survey, must be permanently marked by four or more pegs driven into the roof, with nails in them, or by marks on cross timbers or masonry. Returning to the surface, permanent pegs should be placed at some chains' distance, on each side of the shaft, in a line with the wires, as G and H (see last fig). We then obtain a line on the surface exactly corresponding with the base line of our operation underground. The same process may be adopted, if there is more than one shaft to a mine; but it is not generally desirable to repeat it at more than one shaft. A few hours' labour in getting the fundamental lines permanently fixed and connected, before commencing the survey, is afterwards well repaid.

The angular instrument used for this purpose ought to be the modern improved theodolite (see the foot-note to last article). Three tripods should be provided, and two lamps on stands, fitting on the tripod, of such a height that, when the lamp is replaced by the theodolite, the fulcrum of the axis of the telescope must be of the same height as the top of the wick in the lamp, a tripod with a lamp being placed under the centre of the shaft, at some well-marked station; the second tripod is fixed with the theodolite upon it, as far along the base-line as the light at the bottom of the shaft can be seen. The theodolite is clamped to zero. The third tripod with the other lamp on it, is sent as far forward as the light can be seen from the theodolite. The depth of the top of the wick in the first lamp below the top of the shaft having been ascertained, we carry on a series. of levelling with the vertical arc of the theodolite all through the mine, at the same time as the horizontal angles and the measurement of the lines are taken. The telescope of the theodolite being directed to the top of the wick of the first lamp, the angle of elevation or depression is read. The lower limb being then clamped, and the upper relaxed, the horizontal angle is then read to the second lamp, and at the same time its angle of elevation or depression is read. The distance having been carefully measured, the first tripod is taken up, and carried forward beyond the third tripod, a lamp is placed on the second tripod, and the theodolite on the third tripod, when the observation of the angles are repeated as before.

(22.) The leading feature of Mr. A. Beauland's plan (see "Mining Surveys, Institute of Mining Engineers, Newcastle-upon-Tyne") consists in a method of fixing a bearing, or meridian line at the bottom of the pit, the direction of which is determined, either with reference to the true meridian, or with respect to some line arbitrarily fixed on the surface, as PQ, fig. to Art. 20. By this means the underground survey can be commenced, and carried forward to any extent, by means of the theodolite, and is properly connected with the surface, the whole process being effected without the aid of the magnetic compass.

This method is Mr. B.'s own invention, or, at least, he is not aware that the idea has ever been carried out before, or has ever occurred to any one else, though of course it is quite possible that he may not be the first person who has thought of such a plan.

The process is effected by means of a powerful transit instrument, mounted in the line of the shaft, either at the top or bottom as may be most convenient. For simplicity, suppose the instrument to be at the top of the shaft. It is fixed and properly adjusted on a very firm support, which must be so arranged as not to interrupt the view of the telescope, when pointed vertically down the shaft.

Two marks are then fixed at the bottom of the pit, as nearly as may be in the same vertical plane as the transit, so that each of them can be seen through the telescope, and appear in the centre of the field of view. These marks are rendered visible by the light of a strong lamp reflected upwards, and are likewise so arranged that both can be seen by a theodolite placed at the bottom in a horizontal line with them. They are made as small as will allow of their being observed by the transit at the top, and are of such form that they can be bisected by the wires with great precision, the marks being as far apart as possible.

If now, on pointing the instrument downwards, each of the marks be exactly bisected by the middle wire, it is evident that the horizontal line, in which the marks are placed, coincides with the vertical plane of the instrument, and is, therefore, parallel to the position of the telescope when pointed horizontally. In this case, therefore, we have two lines, one at the top of the shaft, represented by the optical axis of the telescope when pointed horizontally the other the line joining the centres of the two illuminated marks at the bottom, and the bearing of the instrument being determined, either with respect to the meridian, or to some determinate line, which can be connected with the surface survey, that of the line of direction of the marks below is ascertained at the same time.

This, however, is on the supposition, that each of the marks is seen precisely in the centre of the telescope. If this condition is not exactly fulfilled, the marks being a little out of the centre of the field of view, the apparent distance of each mark from the middle wire is accurately measured by a micrometer, or some other means, and from these distances, the angular deviation of the line of the marks from the plane of the instrument is determined by calculation. Having found the amount of this deviation, the bearing of the line of marks is at once deduced from that of the instrument, and the connection between the surface and underground survey, made as in the former case.

It is necessary, in order to complete the process, that permanent marks should be fixed above and below, the marks above ground being set out in some given direction, with respect to the plane of the telescope; those below, with respect to the illuminated marks, which, as well as the instrument, must be removed from their places in the line of the shaft, before the colliery can resume working.

Wherever the nature of the ground, or erections on the surface, admit of it, marks may be placed at once in the direction of the instrument above being set out in any convenient positions, coinciding with the middle wire of the telescope. These permanent marks should of course be placed so that one of them can be seen from the other, it is also desirable to have them conveniently placed for the commencement of the surface survey.

Where, however, it is not practicable to set out a line in the direction of the transit, owing to obstructions, some other direction must be taken, one mark being fixed in the line of the instrument, and the other at any point at a convenient distance, and visible from the first. The direction of the permanent line will, of course, be determined with respect to that of the transit, by setting up the theodolite at the nearer station, and measuring the angle between the direction of the transit and that of the further station.

The permanent marks fixed at the bottom of the pit are fixed in like manner, and their direction determined from that of the illuminated marks, by the aid of the theodolite, which is placed at some point near the shaft, in the line of the illuminated marks, and from which a more distant point can be seen. A permanent mark is then fixed at the place occupied by the theodolite, and another at the more distant point referred to, which may be chosen convenient for the commencement of the underground survey.

Mr. A. B. has thus endeavoured to explain, somewhat briefly, but he trusts with sufficient distinctness, the method by which the underground survey may be connected with the surface. It will scarcely be necessary for him to observe, that the whole process is one requiring great care, and an intimate acquaintance with the use and manipulation of the instruments, such as can scarcely be acquired without considerable expense. With proper management, however, and a transit of sufficient size and power, he believes the bearing may generally be fixed at the bottom of the pit without any error exceeding one minute of an arc, a degree of precision amply sufficient for all practical purposes.

On plotting subterraneous surveys.

(23.) Plotting may be divided into two kinds: *The first kind*, the communicating of bearings and distances of a subterraneous survey to paper, for the purpose of planning the same; *the second kind*, the manner of running on the surface of the earth the different bearings or angles and distances, in the same order as they were taken underground in the survey. In the first mode, the protractor, for setting off the angles contained in each bearing, and a scale of chains and links, for transferring the distances, are requisite; and in the second mode, the circumferentor or theodolite, and Gunter's chain. Observe, in running off the bearings on the surface, that the same instrument be made use of as in the subterraneous survey; and also let the same end of the needle, when used, determine the angles of the bearings as determined them under-ground. This last precaution is not necessary when the magnetic needle is not used.

(24.) Let ABCD represent a protractor, which is a circular rim of brass, and E its centre, of about 9 inches diameter, divided into degrees, and each degree in quarters of a degree, commencing from the north and south points A and B and numbered up to 90° at C and D. Also *abc* represents a semicircular protractor, which for many purposes, is more commodious than the circular one, *ab* representing the meridian, and *e* its centre. These instruments are manufactured by Messrs. Elliott Brothers, 30, Strand, London.

FIG. 34.

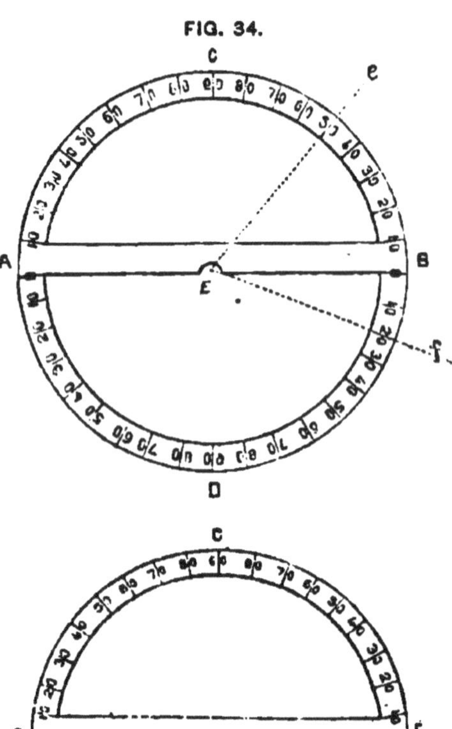

In using this instrument in plotting bearings, the meridian line AB or *ab*, must be applied to the assumed meridian line drawn on paper; and if a line E*e* is drawn from the centre E, through the 50th degree or division from B to C, supposing AB the meridian, A the north, and B the south; then E*e* will be N 50° W (see theorem 17), and the line E*f* passing through the 20th degree or division, will be N 20° E.

(25.) Suppose the following bearings and distances to be plotted on paper—

AB, N. 45° W.	. . .	10 chains.
BC, N. 10° E.	. . .	7 „
CD, S. 50° E.	. . .	6 „

Proceed thus:—

Draw the meridian line NS on the paper where the work has to be plotted, N for north, and S for south; then fix on any place on that meridian line for the commencement of the work, as at the pit A; apply the meridian line AB of the protractor on the assumed meridian, with its centre E on A; let *nesw* represent the protractor, *n* corresponding with N the north, and *s* with S the south,—*e* will represent the east, and *w* the west; then draw the line AB from the centre of the protractor at A through the 45th degree from *n* towards *w*, or west, and it will represent N 45° W: Also from the scale of chains take 10 with the compasses, and setting the same from A to B, and AB will represent the first bearing and distance N 45° W, 10 chains: by the assistance of a parallel ruler, or any other method, draw the second meridian line *ns*, through B, parallel to that drawn through A; apply the protractor

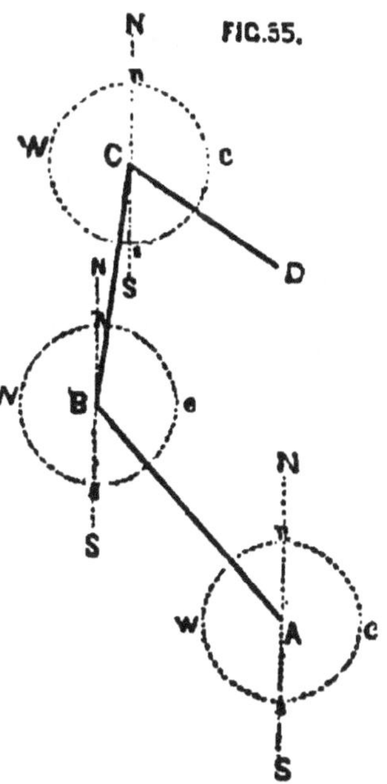

as before directed, with its centre on the point B; draw the line BC from the centre of the protractor at B, through the 10th degree from *n* towards *e*, and it will represent N 10° E: Then from the scale of chains take 7 with the compasses, setting the same from B to C, and BC will represent the second bearing and distance; then draw the third parallel line *ns* through C; apply the protractor as before, with its centre on C; draw the line CD through the 50th degree from *s* towards *e*, and it will represent S 50° E: Then take 6 chains from the scale, setting the same from C to D, and CD will represent the third bearing and distance;—and the whole will be plotted.

Note.—The student ought now to lay down on paper the several surveys, commencing at Art. 12, not only by the old method of bearings, taken by the magnetic needle, but also by the modern and more accurate methods, given in Arts. 20, 21, and 22 (the methods given in the several articles not differing materially except at the commencement of the surveys), that he may thus acquire a skilful and ready method of performing this important part of his profession. See the following article.

(25*a*.) Let the following angles and distances, taken in a coal-mine, be laid down on paper (see fig. to Art. 20).

Let NS be the true meridian, obtained by making proper allowance for the magnetic variation; and let the following distances be measured, and angles be taken in a coal-mine as below:—

DISTANCES.	ANGLES.
AB = 3·12 chains.	NAB = 64° 39′
BC = 4·96 ,,	ABC = 118° 19′
CD = 2·89 ,,	BCD = 79° 15′
DE = 4·17 ,,	CDE = 61° 5′
EF = 6·02 ,,	DEF = 158° 57′

Draw the meridian line NS, N representing the north point, and let A in the line NS be the pit where the work is to commence; lay off from the meridian line NS by the protractor, the angle NAB=64° 39′, in the manner already directed; and from a scale of equal parts, lay off the distance AB=3·12 chains, and extend the line, if necessary; next apply the line AB of the protractor on the line AB on the plan, the centre E of the protractor being applied to the angular point B: then lay off the angle ABC=118° 19′; and the distance BC=4·96 chains; apply the protractor to the line BC, as before directed, lay off the angle BCD= 79° 15′, and the distance CD=2·89 chains: lay off successively the angles CDE=61° 5′ and DEF=158° 57′, and the distances DE=4·17 and EF=6·02 chains; and the work will be completed, being a correct representation of the survey made in the mine.

Next plot the surveys, given in Arts. 17, 18, and 19, by laying off the successive angles, as directed in Arts. 20 and 21, the bearings being previously reduced to the angles, which every two successive distances make with one another by Art. 6.

Note.—In the second column of the survey-book to Art. 17, the angles NAB, ABC, &c., must be entered, as shown in this article.

Suppose the following subterraneous survey is to be plotted by the application of the T square:—

N. 54° W. . . .	10 chains.
S. 42° W. . . .	7 ,,
N. 30° W. . . .	6 ,,

(26.) On the drawing-board, or table ABCD, fix the paper *abcd*, on which the survey is to be plotted, and let SN represent the T square applied thereon, which also represents the magnetic meridian (N the north and S the south). Fix upon the point *f* for the commencement of the work; apply the straight edge of the semicircular protractor *sn* against the arm of the T square NS, with its centre on the point *f*; then draw the line *fg* through the 54th degree of the protractor from north to west, setting off the distance 10 chains from *f* to *g*: Then *fg* is the first bearing and distance N 54° W 10 chains. Remove the T square along the line AC until its arm SN meets the point *g*, where it represents the magnetic meridian; then apply the protractor as before-directed, with its centre on *g*; draw the line *gh* through the 42nd degree from south to west, setting off the distance 7 chains from *g* to *h*: and *gh* is the second bearing and distance S 42° W, 7 chains. Remove again the T square until its arm SN meets the point *h*, representing there the magnetic meridian; then apply the protractor with its centre on *h*; draw the line *hk* through the 30th degree from north to west, setting off the distance 6 chains from *h* to *k*: And *hk* is the third and last bearing and distance N 30° W, 6 chains. The work being finished, take the paper off the board.

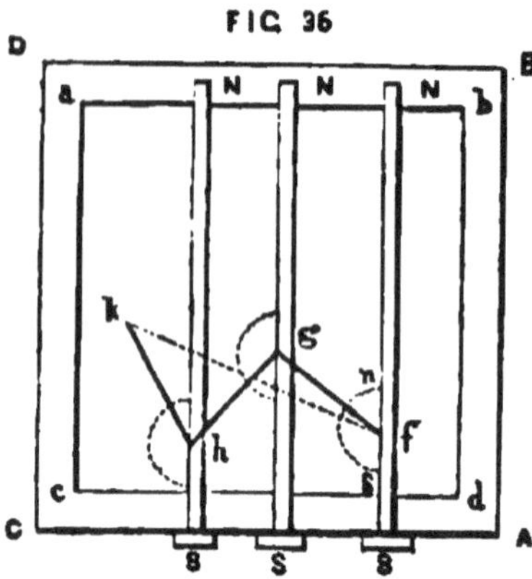

(27.) In the following subterraneous survey *fghk* (see Fig. to last Article), I wish to know, by one single bearing and distance, the situation of *k* from *f*?

fg,	N. 54° W. . . .	10 chains.
gh,	S. 42° W. . . .	7 ,,
hk,	N. 30° W. . . .	6 ,,

Protract the survey on the paper fixed to the drawing-board, as before-directed; then draw a line from *f* to *k*; move the arm of the T square until it touches *f*, forming therewith the magnetic meridian; then apply the protractor with its centre at *f*, observing what division or degree the line *fk* cuts which will be found to be the 71st nearly, which is the magnitude of the angle N*fk*: Measure the distance to *k* from *f* by the same scale as the work was plotted from, which distauce is found to be 16·70 chains; then from rule, Art. 4, the bearing of *k* from *f* will be found to be N 71° W, and its distance 16·70 chains.

(28.) In the following survey of the subterraneous working ABCDF (see Fig. to Art. 16), driven from the pit A towards G, I wish to know the bearing that the workmen must proceed in from F to hit the pit G, and likewise the distance between F. and G?

Plot the survey from the given data in Art. 16, by the use of the T square; also by laying off the several angles, as directed in Arts. 20 and 21, which will verify the survey.

Then the bearing of the pit G from F, from rule, Art. 4, will be S 65° 30′ E. Measure the length of the line FG by the same scale of equal parts as the work was protracted from—which is found to be 8·60 chains; hence the bearing and distance of the subterraneous working from F, to hit the pit G, must be S 65° 30′ E 8·60 chains.

ANOTHER METHOD.

Which may be thought more eligible than the preceding; for if any error is made in this method of plotting, it only affects the particular part where it occurs, and is not carried throughout the remaining part of the work, as in the other methods already described.

(29.) Suppose the following survey to be plotted according to this method:—

	Chains.
AB, S. 36° E.	7·00
BC, S. 42° W.	4·00
CD, S. 75° W.	10·00
DF, N. 42° W.	7·50

Prepare the survey by taking the northing, southing, easting and westing of all the bearings therein (see Art 10, ex. vii.), placing each separately in its respective column, in the following preparatory table: Thus the bearing and distance AB, S 36° E, 7 chains, will, from the traverse tables, contain 5·66 chains of southing, and 4·12 chains of easting;—and so of all the rest.

The next thing is to determine the northing and southing of the bearings conjointly, from A the point of commencement of the survey: Thus let NS represent the magnetic meridian of A, the southing of the bearing AB S 36° E, 7 chains is 5·66 chains A*a*; which place in the 6th column of the preparatory table. The southing of the bearing BC, S 42° W, 4 chains is 2·97 chains *ab*, which, being added to 5·66 chains, makes 8·63 chains A*b* for the southing of the bearings ABC; which place in the 6th column: The southing of the bearing CD S 75° W, 10 chains is 2·59

FIG 37

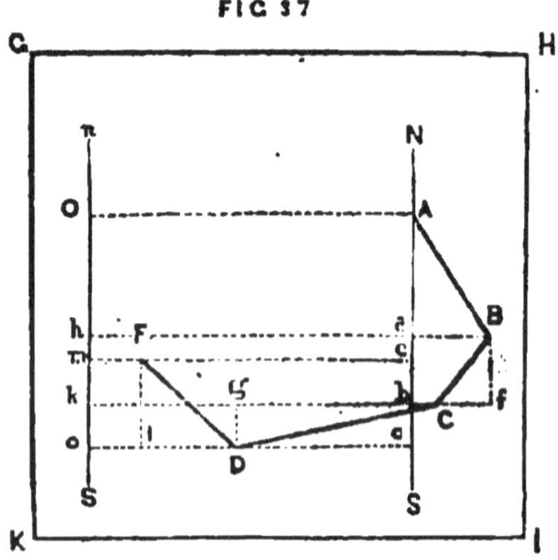

chains *bc*, which, being added to 8·63 chains, makes 11·22 chains A*c* for the southing of the bearings ABCD; which place in the 6th column: The next, DF, N 42° W 7·50 chains, will produce 5·55 chains of northing *ce* from D, which, being subtracted from 11·22 chains, leaves 5·67 chains A*e*, the southing of the bearings ABCDF from the commencement A: Then determine the easting and westing distance of the end of each bearing from the assumed meridian of the point A, or point of commencement. The easting of the bearing AB, which is S 36° E, 7 chains from NS, the assumed meridian, will be found by the traverse tables to be 4·12 chains *a*B; which place in the 7th column of the following table: The westing of the bearing S 42° W, 4 chains from B will be found to be 2·68 chains *f*C, which, taken from the easting *a*B or *bf* 4·12 chains, leaves 1·44 chains, *b*C for the easting of the bearings ABC from NS, the assumed meridian of A: The westing of the bearing S 75° W, 10 chains from C will be found to be 9·66 chains C*g*, from which take 1·44 chains of easting *b*C, leaves *bg* or *c*D 8·22 chains for the westing of the bearings ABCD from NS: The westing of the bearing N 42° W, 7·50 chains from D will be found to be 4·97 chains D*l*, which, being added to 8·22 chains *c*D, makes *cl* or *e*F 13·19 chains for the westing of the bearings ABCDF from NS. Now, to prepare the survey for plotting, the next thing is to assume another meridian, which shall be to the west of the westmost bearing of the survey from NS; and from this second meridian find the easting of the end of each bearing from it (see the 8th column of the table). The greatest westing of the bearing from NS is *e*F, or *cl* 13·19 chains: Suppose, then, this second assumed meridian line to be *ns* 14 chains AO west of the first meridian line NS, place the 14 chains at the top of the 8th column of the following table, which is the distance that the point A is eastward of *ns*: Then 14 chains *ha* + 4·12 chains *a*B = 18·12 chains *h*B, the distance that B is east of *ns*; which place in the 8th column: Then 14 chains + 1·44 chains *b*C = 15·44 chains *k*C, the easting of C from *ns*: Then 14 chains — 8·22 chains *c*D = 5·78 chains *o*D, the easting of D from *ns*: Lastly, 14 chains — 13·19 chains *e*F = 81 links *m*F, the easting of F from *ns*; which, being all entered in the 8th column of the following table, the survey will be prepared for plotting.

PREPARATORY TABLE.

	1.	2.	3.	4.	5.	6.	7.	8.
	Bearings and Distances.	Northing	Southing.	Easting.	Westing.	Northing and southing distance from A.	Easting and westing distance from the meridian of A.	Easting distance of each bearing from the second meridian *ns*.
	Chains.	Chains.	Chains.	Chains.	Chains.	Chains.	Chains.	Chains. OA 14·00 E.
AB	S. 36° E. 7·00	...	A*a* 5·66	*a*B 4·12	...	A*a* 5·66 S.	*a*B 4·12 E.	*h*B 18·12 E.
BC	S. 42° W. 4·00	...	*ab* 2·97	...	*f*C 2·68	A*b* 8·63 S.	*b*C 1·44 E.	*k*C 15·44 E.
CD	S. 75° W. 10·00	...	*bc* 2·59	...	C*g* 9·66	A*c* 11·22 S.	*c*D 8·22 W.	*o*D 5·78 E.
DF	N. 42° W. 7·50	*ce* 5·55	...	...	D*l* 4·97	A*e* 5·67 S.	*e*F 13·19 W.	*m*F 0·81 E.

N.B. The seventh column of the table is only preparatory to the eighth.

In order to plot the survey, fix the paper on the drawing-board or table GHIK; then draw the meridian line *ns* by the application of the T square, *n* representing the north and *s* the south; let O be the point for the commencement of the work, and from the 6th column of the table set off the different southings; 1st, 5·66 chains A*a* from O to *h*, being the southing of the bearing AB; 2dly, 8·63 chains A*b* from O to *k*, the southing of the bearings and distances AB and BC; 3rdly, 11·22 chains A*c* from O to *o*, the southing of the bearings and distances AB, BC, and CD; and 4thly, 5·67 chains A*e* from O to *m*, the southing of the bearing and distances AB, BC, CD, and DF. This being done, apply the T square to the side GK, its arm crossing the meridian line *ns* at right angles; then from the 8th column of the table set off 14 chains of easting from O to A, and A denotes the place of commencement of the survey, or point of departure: Move the T square down the side GK until its arm comes to *h*; then set off 18·12 chains of easting from *h* to B, draw the line AB, and it represents the first bearing and distance; move the T square until the arm comes to *k*, then setting off 15·44 chains of easting from *k* to C, draw the line BC, and it represents the second bearing and distance; move the T square to *o*, then setting off 5·78 chains of easting from *o* to D, draw the line CD, and it represents the third bearing and distance; move the T square to *m*, then setting off 81 links of easting from *m* to F, draw the line DF, and it represents the fourth and last bearing and distance. Then the whole survey will be plotted.

Next plot this survey from the given data by laying off the several angles as directed in Arts. 20 and 21, the bearings being previously reduced to the angles which the successive distances make with one another by Art. 6; also plot the surveys, given in Arts. 30 and 31, in the same manner.

(30.) Suppose the following subterraneous survey to be plotted, beginning at the pit A :—

FIG. 38.

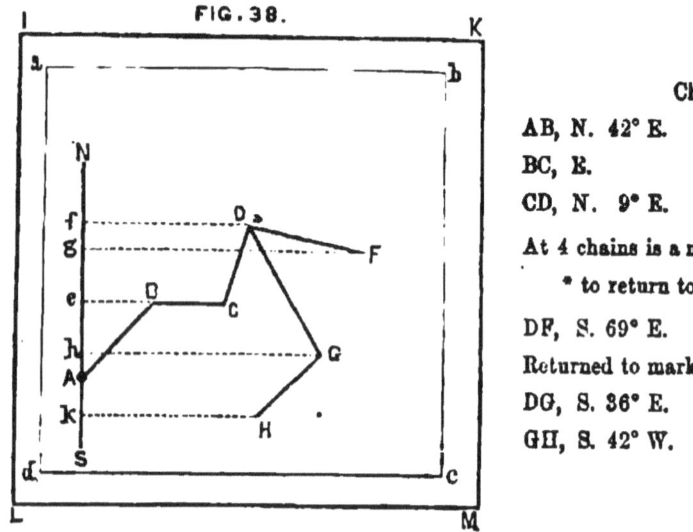

	Chains.
AB, N. 42° E.	5·00
BC, E.	4·00
CD, N. 9° E.	4·00
At 4 chains is a mark * to return to.	
DF, S. 69° E.	5·60
Returned to mark *.	
DG, S. 36° E.	7·00
GH, S. 42° W.	4·00

Prepare the survey for plotting, by taking the northing, southing, easting, and westing of each bearing from the traverse tables.

The northing and easting of the first bearing and distance N 42° E, 5 chains, will be northing 3·72 chains, and easting 3·35 chains (which see in the following preparatory table, together with the northing, southing, easting, and westing of all the others). Then find the northing and southing of the bearings conjointly from the commencement of the survey at the pit A, — which is had from the 2nd and 3rd column of the table: The northing of the first bearing and distance will be found to be 3·72 chains; which place in the 6th column of the table: That of the second bearing and distance will be also 3·72 chains; that of the third, 3·72 chains + 3·95 = 7·67 chains; and so of all the rest. Also take the easting and westing of each bearing and distance from the meridian of the pit A,—which is had from the 4th and 5th column of the table: The easting of the first bearing and distance will be found 3·35 chains; which place in the 7th column of the table: That of the second will be 3·35 + 4 chains = 7·35 chains of easting; and so of all the

PREPARATORY TABLE.

	1.	2.	3.	4.	5.	6.	7.	8.	
	Bearings and Distances.	Northing.	Southing.	Easting.	Westing.	Northing and southing distance from A.	Easting and westing distance from the meridian of A.	Easting distance of each bearing from the meridian of A.	
	Chains.	Chains.	Chains.	Chains.	Chains.	Chains.	Chains.	Chains.	
AB	N. 42° E. 5·00	3·72	...	3·35	...	3·72 N.	3·35 E.	3·35 E.	
BC	E. 4·00	...	...	4·00	...	3·72 N.	7·35 E.	7·35 E.	
CD	N. 9° E. 4·00	3·95	...	0·63	..	7·67 N.	7·98 E.	7·98 E.	mark *
DF	S. 69° E. 5·66	...	2·00	5·21	...	5·67 N.	13·19 E.	13·19 E.	
	Returned to the mark +								
DG	S. 36° E. 7·00	...	5·66	4·12	...	0·01 N.	13·31 E.	17·31 E.	
GH	S. 42° W. 4·00	...	2·97	...	2·68	2·96 S.	14·63 E.	14·63 E.	

rest: As from the 8th column of the table the end of each bearing in the survey will be east of the meridian NS of A; therefore no other need be assumed.

Fix the paper *abcd* on the drawing-board or table IKLM; draw a meridian line NS by the application of the T square, and mark A for the pit, and commencement of the work; then make a mark with the compasses at *e*, on the meridian line NS, 3·72 chains to the north of A (from the 6th column of the table),—which is the northing of the first and second bearing and distance: Make another at *f*, 7·67 chains from A to the north; another at *g*, 5·67 chains; another at *h*, 1 link to the north of A; and another at *k*, 2·96 chains to the south of A: This being done, apply the T square to the side LI, its arm crossing the meridian NS at right angles, and corresponding with *e*; then set off from *e* to B, 3·35 chains of easting to the right (from the 8th column of the table), which is the easting of the first bearing and distance: Draw a line from A to B, and AB represents the first bearing and distance: Also set off from *e* to C, 7·35 chains to the right, and draw the line BC: Remove the arm of the T square to *f*, and set off from *f* to D 7·98 chains to the right, and draw the line CD; there make a mark * to return to: Remove the T square to *g*, and set off from *g* to F 13·19 chains to the right, and draw the line DF: Remove the T square to *h*, and set off from *h* to G 17·31 chains to the right, and draw the line DG from the mark at D: Remove the T square to *k*, and set off from *k* to H 14·63 chains to the right, and draw the line GH;—and ABCDFGH will represent the survey protracted.

(31.) An example showing that an error, committed during the time of plotting the survey after this method, is not communicated to the following part of the work:—

Suppose the subterraneous survey ABCDF is required to be plotted.

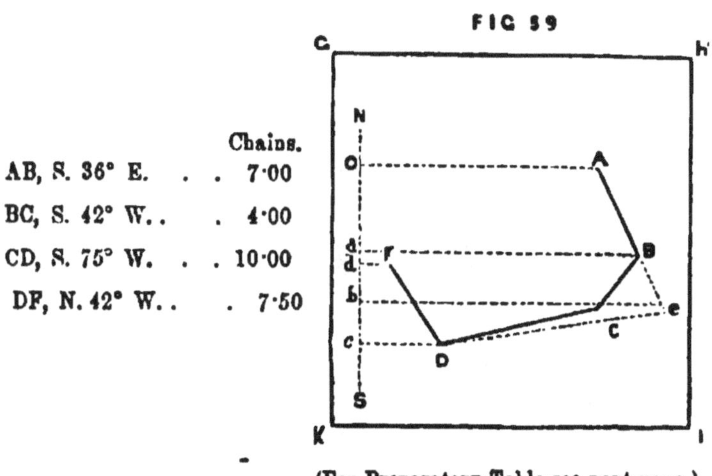

	Chains.
AB, S. 36° E. . . .	7·00
BC, S. 42° W. . .	4·00
CD, S. 75° W. . . .	10·00
DF, N. 42° W. . .	7·50

(For Preparatory Table see next page.)

Fix the paper on the drawing-board GHIK, and draw the assumed meridian NS 14 chains west of the meridian of A, which exceeds the greatest westing in the 6th column; let O be the point thereon for the commencement of the work: Set off, from the 6th column of the table, the different southings from O to *a b c*, and *d* respectively; and from the 8th column of the same table set off the different eastings from O to A, from *a* to B, from *b* to C, from *c* to D, and from *d* to F; and ABCDF will represent the survey truly plotted. Now, suppose the plotter, in laying down the eastings in the 8th column of the table, commits an error, by setting off from the assumed meridian 25·44 chains *be*, instead of 15·44 chains *b*C, then the point C will be removed to *e*, and BC will be represented by B*e*, and CD by *e*D; therefore it appears, from inspecting the figure, that the error will cease at D, and the following bearings, be there ever so many, will be each in the same situation as if no such error had ever existed; which is a peculiar advantage in this mode of plotting.

PREPARATORY TABLE.

	1.	2.	3.	4.	5.	6.	7.	8.
	Bearings and Distances.	Northing.	Southing.	Easting.	Westing.	Northing and southing distance from A.	Easting and westing distance from the meridian of A.	Easting distance of each bearing from the assumed meridian NS.
	Chains.	Chains.	Chains.	Chains.	Chains.	Chains.	Chains.	Chains.
A								14·00 E.
AB	S. 36° E. 7·00	...	5·66	4·12	...	5·66 S.	4·12 E.	18·12 E.
BC	S. 42° W. 4·00	...	2·97	...	2·68	8·63 S.	1·44 E.	15·44 E.
CD	S. 75° W. 10·00	...	2·59	...	9·66	11·22 S.	8·22 W.	5·78 E.
DF	N. 42° W. 7·50	5·55	...	...	4·97	5·67 S.	13·19 W.	0·81 E.

The manner of reducing any number of bearings and distances into one bearing and distance.

(32.) The practical miner will frequently find it necessary to have recourse to this mode of reduction in the plotting of subterraneous surveys on the surface, for the purpose of determining their extent. As when the circumferentor is the only instrument used in such works, which in windy weather is both troublesome and fallacious, therefore, if the whole survey can be reduced to one bearing and distance, or to such a number as may be thought necessary, the labour of protracting will be proportionally reduced, and the work more to be depended on.

Suppose the following subterraneous survey ABCDF, to be reduced to one single bearing and distance from A to F:—

	Chains.
AB, S. 36° E.	7·00
BC, S. 42° W.	4·00
CD, S. 75° W.	10·00
DF, N. 42° W.	7·50

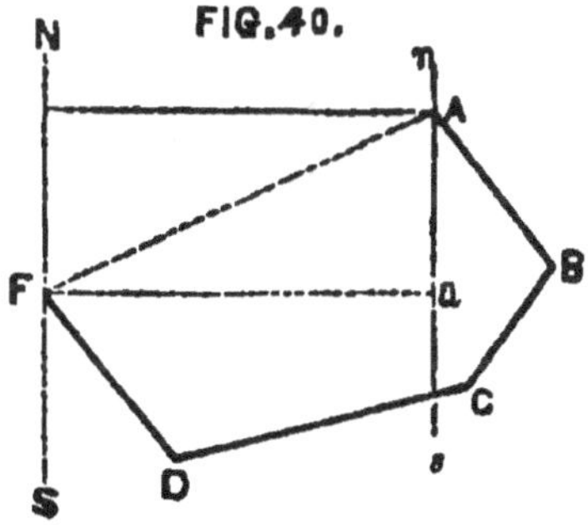

THE PREPARATORY TABLE.

	Northing.	Southing.	Easting.	Westing.
Chains.	Chains.	Chains.	Chains.	Chains.
AB, S. 36° E. 7·00	...	5·66	4·12	
BC, S. 42° W. 4·00	...	2·97	...	2·68
CD, S. 75° W. 10·00	...	2·59	...	9·66
DF, N. 42° W. 7·50	5·55	...	...	4·97
	5·55	11·22	4·12	17·31
		5·55		4·12
	NF or Aa	5·67	NA or Fa	13·19

By taking the northings from the southings, leaves 5·67 chains of southing of F from A; and by taking the eastings from the westings, leaves 13·19 chains of westing of F from the meridian of A, which form the triangle NAF; of which NA 13·19 chains, NF 5·67 chains, and the right ∠ N, are given, to find the side AF and ∠ NAF, which is done by trigonometry, as follows:—

As NA 13·19	1·120245
Is to radius	10·000000
So is NF 5·67	·753583
To tang. ∠ NAF 23° 15′	9·633338
And as sine ∠ A 23° 15′	9·596315
Is to NF 5·67	·753583
So is radius	10·000000
To AF 14·36	1·157268

Or AF may be found thus:—Euclid, b. 1, p. 47, $\sqrt{13{\cdot}19^2 + 5{\cdot}67^2} = 14{\cdot}36$ chains = AF.

Then 90° − 23° 15′ = 66° 45′ ∠ FA*s*. Therefore the bearing and distance of F from A is S 66° 45′ W, 14·36 chains.

Or the bearing and distance may be found instrumentally, if protracted on paper, by applying the protractor to the meridian line *ns*, with its centre on the angular point A, observing the magnitude of the ∠ FA*s*, which will be found 66° 45′; and, from theorem 17, the line AF will bear S 66° 45′ W. The distance may be measured by the scale and compasses.

(33.) In a subterraneous survey ABCDF, commencing at the pit A, I wish to have the direct bearing and distance on the surface of F from A?

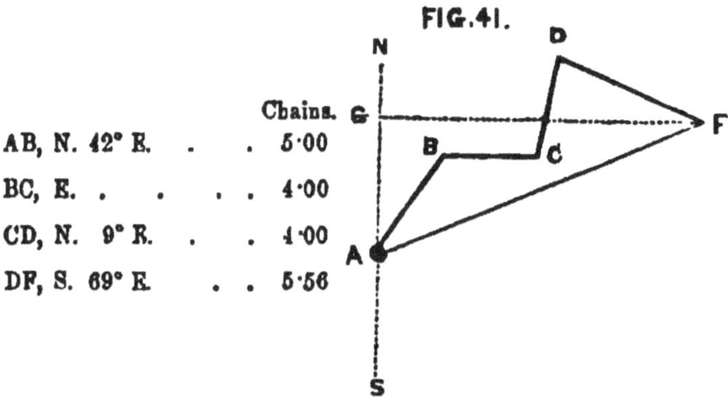

PREPARATORY TABLE.

	Northing.	Southing.	Easting.	Westing.
Chains.	Chains.	Chains.	Chains.	Chains.
N. 42° E. 5·00	3·72	...	3·35	...
E. . . 4·00	...	...	4·00	...
N. 9° E. 4·00	3·95	...	0·63	...
S. 69° E. 5·56	...	2·00	5·21	...
	7·67	2·00	13·19	
	2·00			
	5·67			

Now the point F contains 5 chains 67 links of northing AG, and 13·19 chains of easting GF, from the commencement A of the survey: Therefore construct the triangle AGF, the line NS representing the meridian, N the north, and S the south; AG being given 5·67, and also GF 13·19, and the right ∠ G, to determine AF and the ∠ NAF.

$\sqrt{\overline{13{\cdot}19}^2 + \overline{5{\cdot}67}^2} = 14{\cdot}36$ AF, or the distance of F from A.

As AG 5·67	·753583
Is to radius	10·000000
So is GF 13·19	1·120245
To tang. ∠ A 66° 45′	10·366662

Therefore the bearing and distance of F from A is N 66° 45′ E, 14·36 chains; and if that bearing and distance is run off by a circumferentor and chain, on the surface from A, it will determine the point thereon immediately vertical to the point F in the subterraneous excavation.

Also, if the different bearings and distances ABCDF are protracted on paper, on which the triangle AGF is constructed, beginning at the point A, and making the side AG the meridian, the end of the last bearing and distance DF will coincide with the angular point F of the triangle, if the survey is rightly protracted.

(34.) In the subterraneous survey ABCDFGH, commencing at the pit A, I wish to know the direct bearing and distance of the point D from A, and also the direct bearing and distance of the point H from A, so that a pit may be put down from the surface on each of those points?

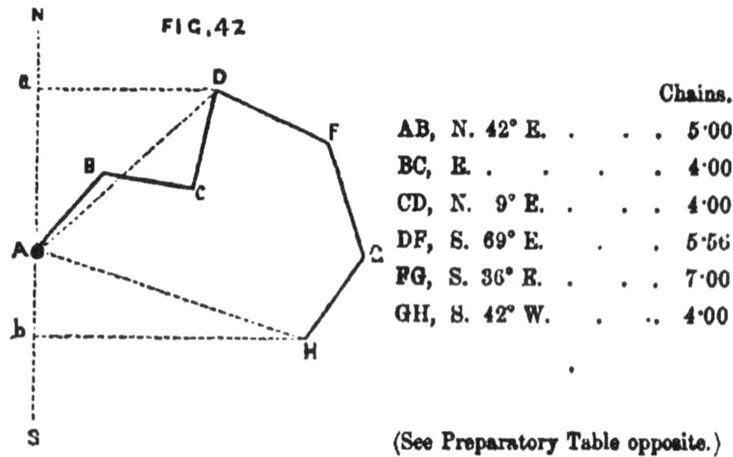

		Chains.
AB,	N. 42° E. . . .	5·00
BC,	E.	4·00
CD,	N. 9° E. . . .	4·00
DF,	S. 69° E. . .	5·56
FG,	S. 36° E. . . .	7·00
GH,	S. 42° W. . ..	4·00

(See Preparatory Table opposite.)

The point D has from A 7·67 chains of northing A*a*, and 7·98 chains of easting *a*D; and the point H has from A 2·96 chains of southing A*b*, and 14·63 chains of easting *b*H.

Construct the triangle A*a*D, and let A*a* represent 7·67 chains of northing, and *a*D 7·98 chains of easting; also construct the triangle A*b*H, and let A*b* represent 2·96 chains of southing, and *b*H 14·63 chains of easting: The side AD

PREPARATORY TABLE.

	Northing.	Southing.	Easting.	Westing.
Chains.	Chains.	Chains.	Chains.	Chains.
N. 42° E. 5·00	3·72	...	3·35	...
E. . . 4·00	...	...	4·00	...
N. 9° E. 4·00	3·95	...	0·63	...
	7·67		7·98	
S. 69° E. 5·56	...	2·00	5·21	...
S. 36° E. 7·00	...	5·66	4·12	2·68
S. 42° W. 4·00	...	2·97	...	...
	7·67	10·63	17·31	2·68
		7·67	2·68	
		2·96	14·63	

and ∠ NAD is required in the former triangle, and the side AH and ∠ SAH in the latter.

Then, as *Aa* 7·67 . . .	·884795
Is to radius	10·000000
So is *a*D 7·98 . . .	·902003
To tang. ∠ A 46° 8′ . .	10·017208

$$\text{And } \sqrt{\overline{7{\cdot}67}^2 + \overline{7{\cdot}98}^2} = 11{\cdot}06 \text{ AD.}$$

Therefore the bearing and distance of a pit from A on the surface, to hit the point D under-ground, will be N 46° 8′ E, 11·06 chains.

Also, as A*b* 2·96 . . .	·471292
Is to radius	10·000000
So is *b*H 14·63 . . .	1·165244
To tang. ∠ A 78° 33′ . .	10·693952

$$\text{And } \sqrt{\overline{14{\cdot}63}^2 + \overline{2{\cdot}96}^2} = 14{\cdot}92 \text{ AH.}$$

Consequently the bearing and distance of a pit from A

on the surface, to hit the point H under-ground, will be S 78° 33′ E, 14·92 chains.

(34a.) In the subterraneous survey ABCDEF, commencing at the pit A, it is required to find the direct bearing and distance of the point F from A, so that a pit may be sunk from the surface to the point F (see Fig. to Art. 20), the required bearing being taken both from the meridian NS, and also from a well defined line GH, passing through the fixed marks G and H and the shaft A, and corresponding to a line in the headway AB, determined in the manner pointed out in Articles 21 and 22.

Let the distances be measured, and the angles be taken by the theodolite as below:—

Distances.	Angles.
AB = 3·12 chains.	
BC = 4·96 ,,	ABC = 118° 34′
CD = 2·89 ,,	BCD = 79° 15′
DE = 4·17 ,,	CDE = 61° 5′
EF = 6·02 ,,	DEF = 158° 57′

Reduce the angles at B, C, D and E to their bearings from GH by Art. 3; then find the northing or southing and the easting or westing from the Traverse Table by Art. 59; then proceed as in Art. 34 to find Ag and Fg; whence by trigonometry, as shown in the last-named Article, the bearing of F from the fixed line GH, and the distance AF, will be readily found.

Next reduce the angles A, B, C, D and F to their bearing from NS, then find the northing or southing and easting or westing from the Traverse Table, and proceed as in Art. 34 to find Am and Fm; whence by trigonometry, as already shown, the bearing of F from the meridian NS and the distance AF will be found.

Or the bearings and distance in both cases may be found from the plan by measuring the angles GAF and NAF with the protractor, and the distance AF by the same scale of equal parts as that with which the plan was laid down. By doing the work by all these methods its accuracy may be further verified.

It would conduce much to the improvement of the student to plot the surveys in the following Articles 35 and 36, by reducing the given bearings to the angles made by every two successive lines in each example, as practice of this kind will impart great facility in the exercise of his profession; and besides, enable him to reason for himself and not on every slight occasion to have recourse to authors.

(35.) In the following subterraneous working ABCDF, beginning at the pit A, I wish to know the bearing and distance of the pit G from F, the bearing and distance of G from A being given:—

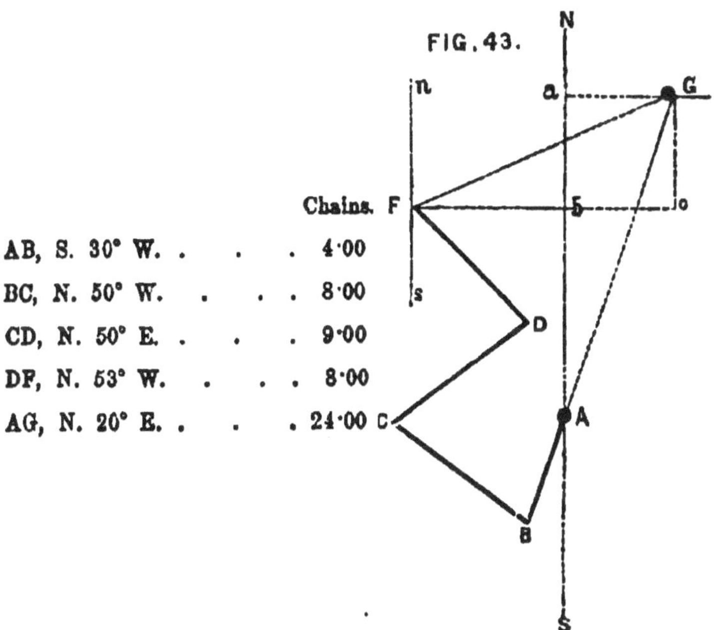

	Chains.
AB, S. 30° W.	4·00
BC, N. 50° W.	8·00
CD, N. 50° E.	9·00
DF, N. 53° W.	8·00
AG, N. 20° E.	24·00

PREPARATORY TABLE.

		Northing.	Southing.	Easting.	Westing.	
	Chains.	Chains.	Chains.	Chains.	Chains.	
S. 30° W.	4·00	...	3·46	...	2·00	
N. 50° W.	8·00	5·14	...	...	6·13	
N. 50° E.	9·00	5·79	...	6·89	...	
N. 53° W.	8·00	4·81	...	...	6·39	
		15·74	3·46	6·89	14·52	
		3·46			6·89	
		12·28	Ab		7·63	bF
N. 20° E.	24·00	22·55	Aa	8·21	aG or bc	
		10·27				

Then A*a* 22·55 chains — A*b* 12·28 chains = 10·27 chains *ba* or *c*G, and F*b* 7·63 chains + *a*G or *bc* 8·21 chains = 15·84 chains F*c*.

The ∠ *n*FG, or bearing of the line FG from the magnetic meridian *ns*, and the length of FG are both wanted; and the sides F*c*, *c*G, and the right ∠ *c* are given to find them.

As F*c* 15·84	1·1997552
Is to radius	10·0000000
So is G*c* 10·27 . . .	1·0115704
To co-tang. ∠ F 57° 3′ . .	9·8118152

Which will be N 57° 3′ E with the magnetic meridian *ns*.

Also $\sqrt{15{\cdot}84^2 + 10{\cdot}27^2} = 18{\cdot}87$ FG.

Therefore the bearing of the pit G from F will be N 57° 3′ E, and the distance 18·87 chains.

Plotting on the surface by the circumferentor, or theodolite.

(36.) In this mode of plotting the bearings and distances are run off on the surface of the earth in the same order as taken in the subterraneous survey. Great care must be taken in running the length of each bearing as nearly horizontal as can be, where the surface is uneven and declining.

—The first two examples show the different modes of commencing the plotting of a survey on the surface, by assuming a point to begin at; and the others following show the manner of avoiding an obstacle, as a house, a lake, or any other thing that interferes with the line of survey.

Let the following subterraneous survey be plotted on the surface, commencing at the centre of the pit A:—

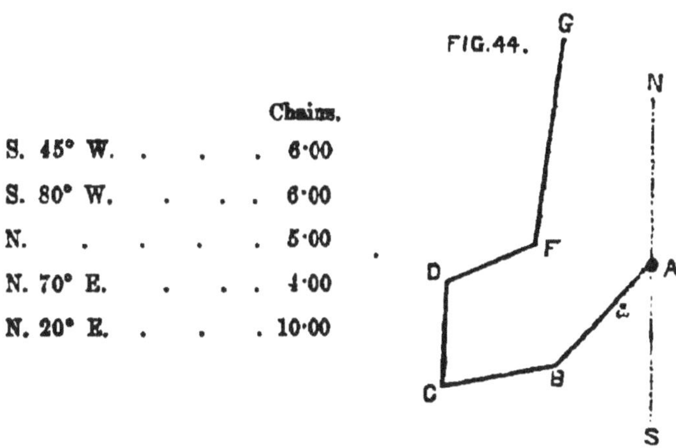

	Chains.
S. 45° W.	6·00
S. 80° W.	6·00
N.	5·00
N. 70° E.	4·00
N. 20° E.	10·00

Fix the instrument as near the pit A as convenience will allow; (observe to keep the same end of the instrument first in the plotting of the survey as was first in making it under-ground; likewise the same end of the needle must determine the bearings in the plotting as determined them under-ground). Suppose *a* the place where the instrument is fixed, which is such a situation that, when the fore-sight is put in the direction of the first bearing, S 45° W, you may, by looking backward from *a*, cut exactly the centre of the pit A, the commencement of the survey,—otherwise the instrument is not placed in a proper situation. (This first point A is obtained by shifting the instrument either to the right or left, until it is in the situation before-mentioned.)—After the proper situation of the commencement of the survey is found, let the assistant take the chain, and running 6 chains from the centre of the pit A, which suppose to extend to B, then AB is the first bearing and distance plotted. Remove the instrument to B, and put the fore-sight in direction of S 80° W, measuring the distance from B to C 6 chains; then BC is the second bearing and distance. Remove again the instrument to C, and put the foresight in direction of due north, measuring the distance from C to D 5 chains; then CD is the third bearing and distance. Remove again the instrument to D, and put the fore-sight in direction of N 70° E, measuring from D to F 4 chains; then DF is the fourth bearing and distance. Lastly, remove the instrument to F, and putting the fore-sight in direction of N 20° E, measure 10 chains from F to G; then FG is the fifth and last bearing and distance. If marks are made at B, C, D, F and G, they will represent on the surface the excavation with all its windings.

(37.) Suppose the following subterraneous survey ABCDF, to be plotted on the surface, commencing at the centre of the pit A:—

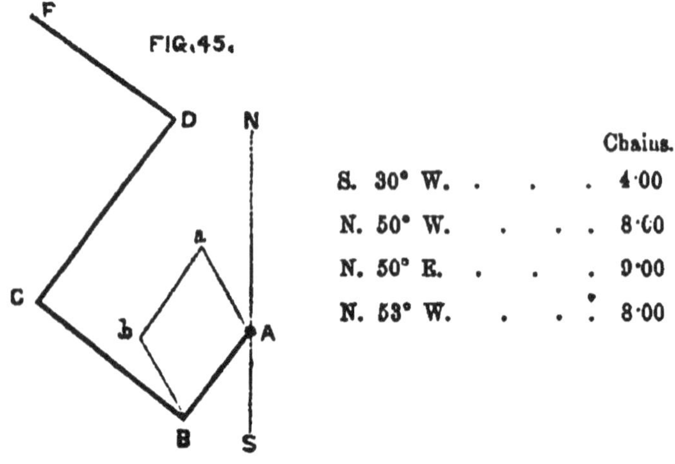

	Chains.
S. 30° W. . . .	4·00
N. 50° W. . . .	8·00
N. 50° E. . . .	9·00
N. 53° W. . . .	8·00

Instead of following the same mode of commencement, as shown in the former example, make any place on the surface the point of commencement, as *a* (the same not being far distant from the pit A), and run off from that assumed point *a* the first bearing and distance, in the same manner as if *a* was the centre of the pit A; which first bearing and distance S 30° W 4 chains suppose to be represented by *ab*. Before the instrument is removed, from *a* take the bearing and distance of the centre of the pit A from *a*, which suppose S 30° E 3 chains *a*A, and insert it in the column of remarks in the survey-book (for fear it should be forgot), as a deflection from the line of the subterraneous survey; which deflection must be accounted for before the whole of the survey is plotted. Now remove the instrument to *b*, and there turn the sights in the direction of S 30° E, running off 3 chains, which let *b*B represent; then the line AB represents the first bearing and distance as if taken from the centre of the pit A (the line of deflection A*a* is now repaid). Remove the instrument from *b* to B, and proceed to run off the second bearing and distance N 50° W 8 chains BC, according to the method described in the last example. Remove the instrument to C, and run off the third bearing and distance N 50° E 9 chains CD. Lastly remove the instrument to D, and run off the fourth bearing and distance N 53° W 8 chains DF. And if marks are put up at BCD and E, they will represent the course of the subterraneous excavation on the surface.

Note.—This survey ought also to be plotted by the new methods, given in Arts 20 and 21, in the manner directed in Art. 34*a*.

To show the manner how to avoid an obstacle that interferes with the line of survey when plotting it on the surface of the earth.

(38.) Suppose the following survey ABCD is to be plotted on the surface, commencing at the centre of the pit A;—

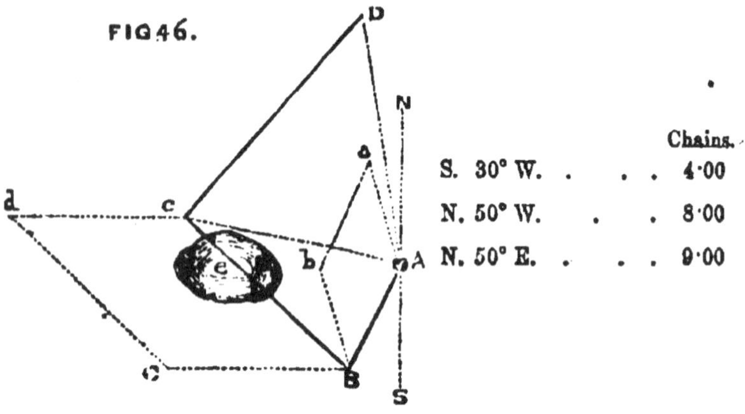

	Chains.
S. 30° W.	4·00
N. 50° W. . . .	8·00
N. 50° E.	9·00

Fix the instrument at the point *a*, as the assumed centre of the pit A, and run off the first bearing and distance from thence, which suppose to extend to *b*; then take the bearing and distance of A from *a* for the deflection, which note down in the column of remarks in the survey-book: suppose it to be S 30° E 3 chains: Then remove the instrument to *b*, and from thence run off S 30° E 8 chains *b*B, and the line AB will be the first bearing and distance as run off from A. Remove the instrument to B, and proceed to plot the remaining part of the survey. Now the next bearing, N 50° W, will be found to run over the lake *e*; therefore, to avoid this obstruction, let the plotter extend the line B*c* to such a distance that, in running the second bearing from *c*, he may avoid the obstruction. Suppose this line B*c* to be due west 6 chains, which being noted down in the survey-book, remove the instrument to *c*, and from thence let the bearing N 50° W 8 chains be run, which suppose it to extend to *d*; then from *d* run off due east 6 chains (being the reverse of B*c*), which suppose to extend to C; then BC will be the second bearing and distance. Remove the instrument to C, and run off the third-bearing and distance, which suppose to extend to D; then CD will represent N 50° E 9 chains, and the whole is plotted.

The obstruction at *e* may be more easily avoided by

laying down the survey on paper, and drawing on it the lines AC AD, which must be measured, and their bearings from NS found; thus the position of the points C and D may be determined; also various other similar methods will readily suggest themselves to the student when the obstructions are even more formidable than that at *e*.

In plotting a survey, either on paper or on the surface of the earth, it matters not whether we begin with the first or last bearing, the ending will be the same.

(39.) Thus, suppose the subterraneous bearings and distances are required to be plotted, in order to determine on the surface the situation of the end D from the commencement A:—

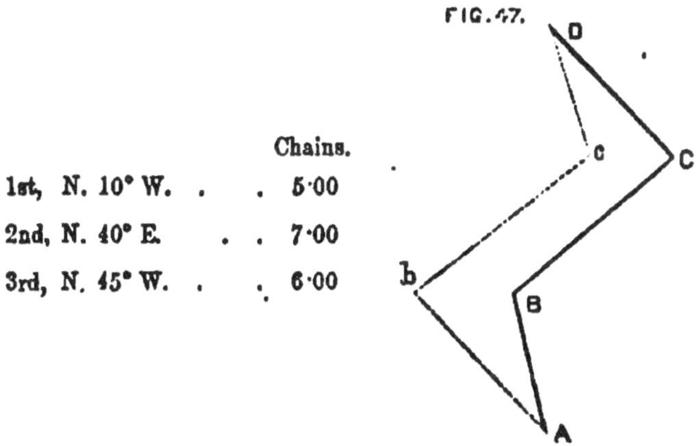

	Chains.
1st, N. 10° W.	5·00
2nd, N. 40° E.	7·00
3rd, N. 45° W.	6·00

Suppose A the point or place of commencement, and run off from thence the first bearing and distance to B, then AB will represent N 10° W 5 chains; and from B run off the second bearing and distance to C, then BC will represent N 40° E 7 chains; and from C run off the third bearing and distance to D, then CD will represent N 45° W 6 chains,—and the whole is plotted in the order of the survey. Now, to plot the same in a manner contrary to the order of the survey, begin at the point A, and run off the third bearing and distance N 45° W 6 chains,—which let Ab represent; run off from b the second bearing and distance N 40° E 7 chains bc; also run off from c the first bearing and distance N 10° W 5 chains cD; which termination will correspond with the point D in the former method, if the work be right.

(40.) In the following subterraneous survey ABCDF, beginning at the pit A, I wish to have the same plotted on the surface, in order to determine the bearing and distance of F from A:—

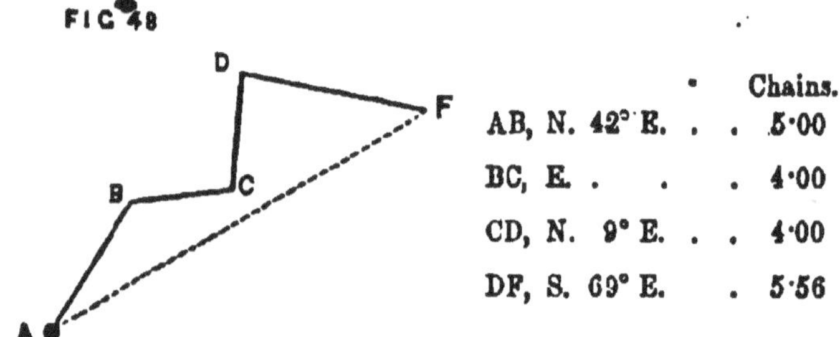

	Chains.
AB, N. 42° E. . .	5·00
BC, E. . . .	4·00
CD, N. 9° E. . .	4·00
DF, S. 69° E. .	5·56

Commence the plotting on the surface according as directed in the former examples; and running off the bearing and distance AB N 42° E 5 chains, remove the instrument to B, and run off the bearing and distance BC due east 4 chains; remove the instrument to C, and run off the bearing and distance CD N 9° E 4 chains; lastly, remove the instrument to D, and run off the bearing and distance DF S 69° E 5·56 chains, and there make a mark; then return with the instrument to the pit A, and take the bearing of the mark at F, which suppose N 64° 44′ E; then measure the distance, which suppose 14·40 chains, which are the bearing and distance required. Or the circumferentor may be fixed at F instead of A, and the bearing of A taken from it; which being reversed (see Art. 5), will become the bearing of F from A, the same as before.

Note.—In many cases, where the surveyor is desirous of plotting the subterraneous survey on the surface, it will be best to make choice of a level piece of ground, sufficiently large to contain the whole, and plot the same thereon, assuming a point of commencement in the most advantageous place.

(41.) In the following survey of the subterraneous working ABCDF, driven from the pit A, I wish to know by what bearing the miner must be conducted from F, the extreme point of the excavation, just to hit the centre of the pit G; and also what is the distance of G from F?

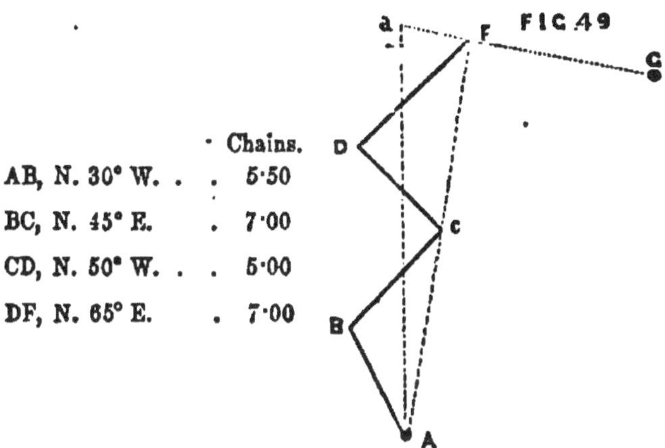

	Chains.
AB, N. 30° W. . .	5·50
BC, N. 45° E. .	7·00
CD, N. 50° W. . .	5·00
DF, N. 65° E. .	7·00

Commence the plotting at the pit A on the surface, as before directed, running off the distances in direction of their respective bearings: When the whole is run off to F, fix the instrument there, and take the bearing of the pit G from it, which suppose S 68° 30′ E; then measure, by the chain, the distance of G from F, which suppose 8·60 chains, the direction and distance required to hit the pit G.

In the foregoing subterraneous survey, commencing at the pit A, I wish to know the bearing and distance on the surface of F from A, without plotting the same?

(42.) Reduce the bearings and distances of the survey to their northing or southing, and easting or westing (see Art. 10, Ex. VII.), in order to obtain the denomination of bearing of F from the pit A. Thus:—

PREPARATORY TABLE.

	Northing.	Southing.	Easting.	Westing.
Chains.	Chains.	Chains.	Chains.	Chains.
N. 30° W. 5·50	4·76	...	..	2·75
N. 45° E. 7·00	4·95	...	4·95	...
N. 50° W. 5·00	3·21	...	...	3·81
N. 65° E. 7·00	2·96	...	6·34	...
	15·88	Aa	11·29	6·56
			6·56	
			4·73	aF

The denomination of bearing of the extreme part of the excavation F from the pit A is 15·88 chains of northing, and 4·73 chains of easting. Therefore,

As Aa 15·88	1·2008505
Is to radius	10·0000000
So is aF 4·73	·6748611
To tang. ∠ aAF 16° 35′ . .	9·4740106

$$\text{And } \sqrt{\overline{15 \cdot 88}^2 + \overline{4 \cdot 73}^2} = 16 \cdot 56 = \text{AF}.$$

Now fix the instrument at the pit A, and run off from thence the bearing and distance N 16° 35′ E 16·56 chains, and the situation of F, with respect to the pit A, will be had on the surface.

(43.) In the workings of the pit A, see fig. 30 to Art. 17, I wish to know how far each bord or excavation *op*G*qrst* is distant from the boundary *cdfgmn*?

To obtain what is required, fix the circumferentor at the pit A, and survey in direction of A*a*G, VF, and *b*H, which are the excavations next the boundary,—measuring the distance that each bord *op*G*qrst* is driven towards the boundary from the headways VF and *b*H, entering them, according to the following form, in the survey-book.

SURVEY-BOOK.

Bearings.	Remarks to Left.	Distance.	Remarks to Right.	
		Chains.		
S. 80° W.	. . . . .	1·60	. . . . .	Aa
S. 70° W.	. . . . .	1·80	. . . . .	aG
		0·80	A headways b, and a chalk mark + to return to.	
	A headways V, and a chalk mark + to return to.	1·20		
	Returned to . .		mark + at V.	
S. 10° W.	. . . . .	2·40	. . . . .	VF
		0·80	Bord p, 1·30 chains towards the boundary.	
		1·60	Bord o, 1 chain towards ditto.	
	Returned to . .	...	mark + at b.	
N. 1° W.	. . . . .	5·00	. . . . .	bH
	Bord q, 90 links towards the boundary	0·80		
	Bord r, 60 links towards the boundary	1·70		
	Bord s, 60 links towards the boundary	2·55		
	Bord t, 55 links towards the boundary	3·40		

Note.—The student must recollect to enter in this and all other survey-books in the first column the angles which every two successive lines in the survey make with one another when the new methods, given in Arts. 20 and 21, are used; besides, not only this survey, but also the following one, ought to be done without the use of the magnet.

Now the survey underground being finished, fix the instrument at the pit A, on the surface, and run off the bearing and distance therefrom, in the order as taken underground,—the first S 80° W. 1·60 chains A*a*: Remove the instrument to *a*, and run off the next bearing and distance S 70° W 1·80 chains *a*G; at 80 links make a mark on the surface, as represented by *b*; also at 1·20 chains make another, as represented by V; and at G make another: Then with the chain measure the distance G*f*, which suppose 1·30 chains, which is the distance the excavation G is short of the boundary,—which must be recorded in the miner's book of memorandums. Return with the instrument to the mark made on the surface at V, and run off S 10° W 2·40 chains VF; at 80 links, in direction from V to F, run off the bord *p* 1·30 chains to the right, perpendicular to the headways VF, and there make a mark at *p*: Then measure the distance *pe*, which suppose 1 chain, which is the distance of the bord *p* from the boundary,—which must be recorded. Also at 1·60 chains run off the bord *o* 1 chain to the right, similar to the former, and make a mark at *o*: Then measure the distance *od*, which suppose 70 links, which is the distance of the bord *o* from the boundary,—which must also be recorded. Return with the instrument to the mark made on the surface at *b*, and run off N 1° W 5 chains *b*H, and there make a mark: Then measure the distance H*n*, which suppose 80 links, which is the distance of the headways *b*H from the boundary. At 80 links, in the direction from *b* to H, run off the bord *q* 90 links to the left, perpendicular to the headways *b*H, and there make a mark at *q*: Then measure the distance *qg*, which suppose 70 links, which is the distance of the bord *q* from the boundary. From 1·70 chains run off the bord *r* 60 links to the left, and make a mark at *r*: Then measure the distance *rh*, which suppose 1·10 chains, which will be the distance of the bord *r* from the boundary. From 2·55 chains run off the bord *s* 60 links to the left, and make a mark at *s*: Then measure the distance *sk*, which suppose 1·50 chains, which will be the distance of the bord *s* from the boundary. From 3·40 chains run off the bord *t* 55 links to the left, and make a mark at *t*: Then measure the distance *tm*, which suppose 1·80 chains, which will be the distance of the bord *t* from the boundary,—and the whole will be finished.

(44.) In the following subterraneous survey ABCDF, commencing at the pit A, I wish to know the bearing and distance of F from A :—

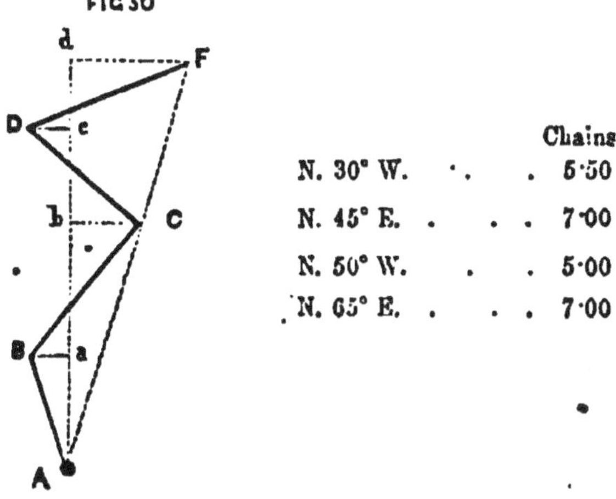

	Chains.
N. 30° W. . . .	5·50
N. 45° E. . . .	7·00
N. 50° W. . .	5·00
N. 65° E. . . .	7·00

PREPARATORY TABLE.

1.	2.	3.	4.	5.	6.	7.
Bearings and Distances.	North-ing.	South-ing.	East-ing.	West-ing.	Northing and southing distance from A.	Easting and westing distance from the meridian of A.
Chains.	Chains.	Chains.	Chains.	Chains.	Chains.	Chains.
N. 30° W. 5·50	4·76	...	...	2·75	4·76 N.	2·75 W.
N. 45° E. 7·00	4·95	...	4·95	...	9·71 N.	2·20 E.
N. 50° W. 5·00	3·21	...	...	3·81	12·92 N.	1·61 W.
N. 65° E. 7·00	2·96	...	6·34	...	15·88 N.	4·73 E.

Fix the instrument at A, and run off the line A*d* 15 chains 88 links on the magnetic meridian of A (from the 6th column of the table), for the northing of F from A; and also 4 chains 73 links *d*F (from the 7th column of the table), for the casting of F from the magnetic meridian of A: Then at F fix up a mark, and take its bearing and distance from A, which suppose N 16° 35′ E 16·56 chains,—which is the bearing and distance required.

This mode of plotting will be tedious, and liable to error, particularly where the surface is uneven.

The manner of making a survey where the subterraneous excavation declines from the horizon.

(45.) In making surveys where the distances measured are not horizontal, but rising or falling, or both, it will be necessary for the surveyor to reduce all his measurements to horizontal distances, which may be obtained by taking the angle that each separate distance makes with the horizon, noting the same down opposite its respective bearing, in a column made for that purpose in the survey-book.

FIG. 51

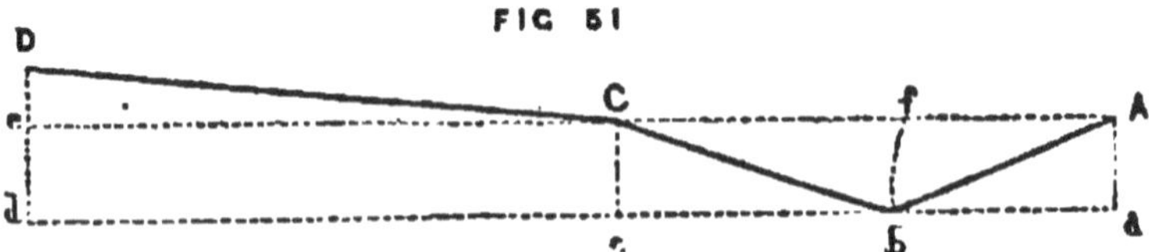

Suppose A*e* and *ad* to be lines parallel to the horizon, and A*b*CD is the undulating excavation which is to be surveyed, commencing at A; let the bearing and distance taken in such a situation as that of A*b* to be N 10° W 5 chains, and the angle *f*A*b* which such excavation makes with the horizon to be 30°; and another in such a situation as that of *b*C N 20° W 6 chains, and the angle C*bc* which it makes with the horizon to be 20°; also another in the situation of CD N 20° E 12 chains, and the angle DC*e* which it makes with the horizon to be 10°;—which bearings, distances, &c., must be inserted in the following survey book:—Thus, the first column containing the bearings and declining distances, the second column the magnitude of the angle that each bearing forms with the horizon, and the third the declining distance of each bearing reduced by the traverse tables to horizontal distance. This third column may be made at the surveyor's leisure, but previous to its being plotted.

SURVEY-BOOK.

1.		2.	3.	
Rising and falling distances.		Angle that each bearing forms with the horizon.	The horizontal distance of each bearing.	
	Chains.		Chains.	
N. 10° W.	5·00	30°	4·33	*Af* or *ab*
N. 20° W.	6·00	20°	5·64	*bc* or *Cf*
N. 20° E.	12·00	10°	11·82	C . or *cd*

I shall protract the survey first without reducing the declining measurements to horizontal distances, from the first column of the foregoing survey-book; and, secondly, by the same, reduced to horizontal distances, taken from the third column,—in order to show the error arising from the protracting of declining or hypothenusal distances.

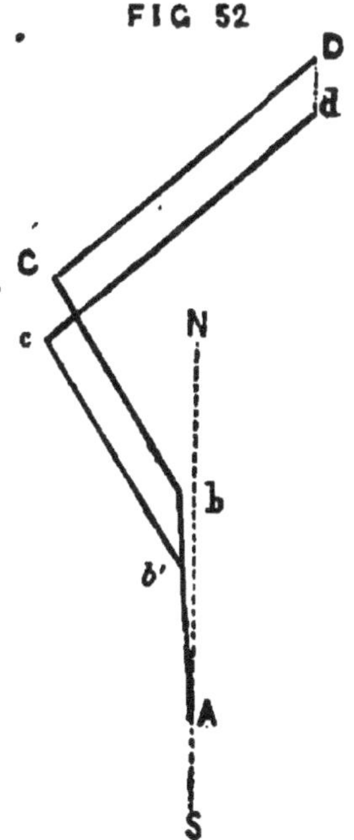

Without reducing the declining distances.—Let A*b* represent N 10° W 5 chains, *b*C N 20° W 6 chains, and CD N 20° E 12 chains; then A*b*CD will represent the survey protracted according to the first column of the survey-book.

Where the declining lengths of each bearing are reduced to horizontal dis-

tance.—Let *Ab′* represent N 10° W 4·33 chains (from the 3rd column of the survey-book), *b′c* N 20° W 5·64 chains, and *cd* N 20° E 11·82 chains; then *Ab′cd* will represent the true protraction, and A*b*CD the false one,—and *d*D will be the amount of the error.

As it is common among practical miners, when plotting their surveys, to add a number of bearings and distances together, taking the mean sum of the degrees contained in the bearings so added for the common bearing of the whole, when they are all on the same side of the same meridian,—and the sum of the lengths of all the bearings for the length of the whole,—I shall therefore show the errors which result from such practices.

(46.) Suppose AB N 30° W 10 chains, and BC N 50° W 20 chains to be plotted.

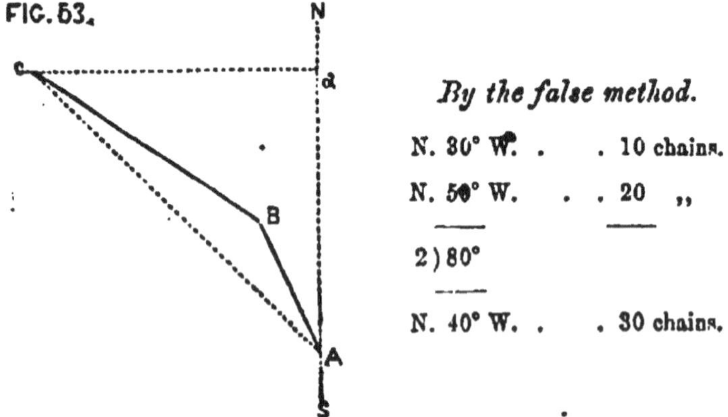

Now it appears that N 40° W 30 chains will be the bearing and distance equal to both, by this method.

By the true method.

		Northing.	Westing.	
	Chains.	Chains.	Chains.	
N. 30° W.	10·00	8·66	5·00	
N. 50° W.	20·00	12·86	15·32	
	Aa	21·52	20·32	aC

As 21·52	1·3328420
Is to radius	10.0000000
So is 20·32	1·3079240
To tang. ∠ aAC 43° 21′	9·9750820

And $\sqrt{21{\cdot}52^2 + 20{\cdot}32^2} = 29{\cdot}59$ chains $=$ AC.

Then N 43° 21′ W 29·59 chains is the true bearing and distance of C from A, instead of N 40° W 30 chains;—and the magnitude of the error will be their difference, *i. e.* 30—29·59=41 links; it is hence presumed that no surveyor will use the false method.

A promiscuous collection of practical questions.

(47.) Example I.—I wish to drive a drift or subterraneous excavation from the point A to hit the pit C, which is on the other side of a river; now I run a line AB by the river side, N 85° E 20 chains long, which from the point A, I found C to bear N 42° E, and also from the point B, I found C to bear N 30° W; I therefore desire to know what will be the length of the excavation or drift AC?

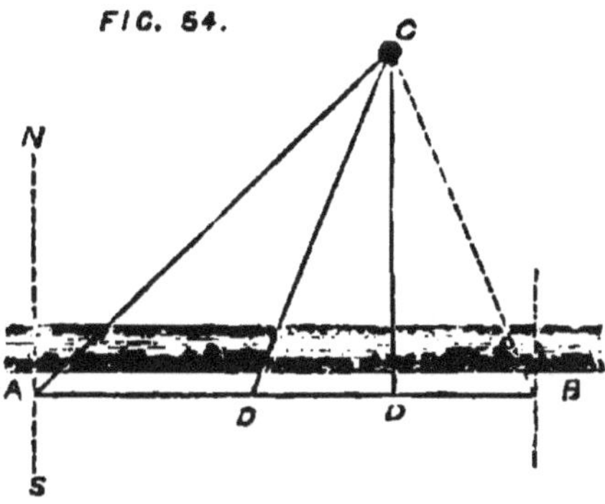

From the rules for reducing bearings into angles, ∠A =43°, ∠B=65°, and ∠C=72°; therefore, by trigonometry, the excavation AC will be 19·05 chains in length.

EXAMPLE II.—There is a pit C (see last fig.), on the other side of a river, to which I wish to drive a drift from a given point; I took the bearing of the pit C from A, which was N 42° E, and after running a line AB by the river side in direction of N 85° E 20 chains, I also took its bearing again from B, which I found to be N 30° W; now I demand to know under what bearing I must set off a drift from a point D, 8 chains from A, along the line AB, so that I may hit C,—and also what will be the length of the drift DC?

∠A=43°, ∠B=65°. The line DC, which is the direction of the drift or excavation, will be found to form an angle with the line DB of 65° 33′, and the line DB forms an angle of 85° to the left of the north magnetic meridian; therefore, from the rule for reducing angles into bearings (Art. 3), the drift DC will bear N 19° 27′ E, and its length will be 14·29 chains.

EXAMPLE III.—There is an inaccessible point C (see fig. to Ex. I.), to which a drift is to be driven underground; now the angle A is found to be 43°, the angle B=65°, and the length of the drift AB=20 chains: how far from A must a drift be set off to arrive at C by the shortest distance possible?

The shortest distance between AB and the point C is a line perpendicular to AB, let fall from the point C.

The point, from which the drift DC must be driven, by the shortest distance possible, will be 13·94 chains from A, as required.

EXAMPLE IV.—I made a survey along the side of a hill from A to B, under the following bearings and distances viz., A*a* N 75° E 20 chains, *ab* S 80° E 19 chains, *bc* N 73° E 15·60 chains, and *c*B S 71° E 18·65 chains; now I wish to make a straight tunnel from A to B; therefore I demand to know under what bearing it must be conducted, and what will be its length?

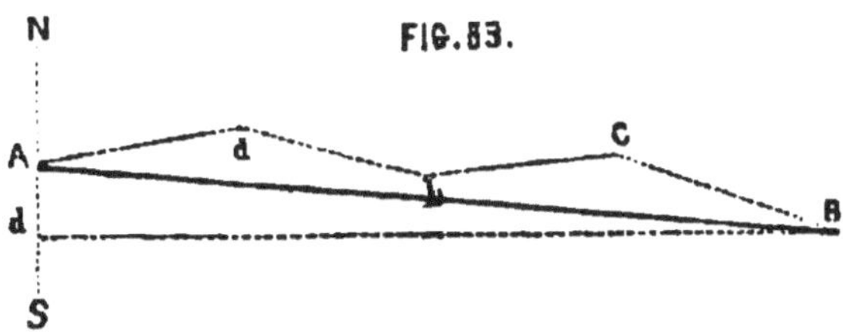

From the traverse tables, the point B will have 7·63 chains of southing *Ad*, and 65·17 chains of easting *d*B from A; and the angle SAB = 83° 43′, which is the angle that the line AB makes with the south meridian,—and AB being the direction of the tunnel, therefore it must be conducted from A to B under the bearing of S 83° 43′ E, and its length will be 65·61 chains.

Example V.—There is a vein of lead ore AB, which I find forms an angle CAB of 82° with the horizon; now I wish to know how deep my shaft DB must be sunk before I cut the vein, if I set it off at the distance of 80 yards from A to D at the surface?

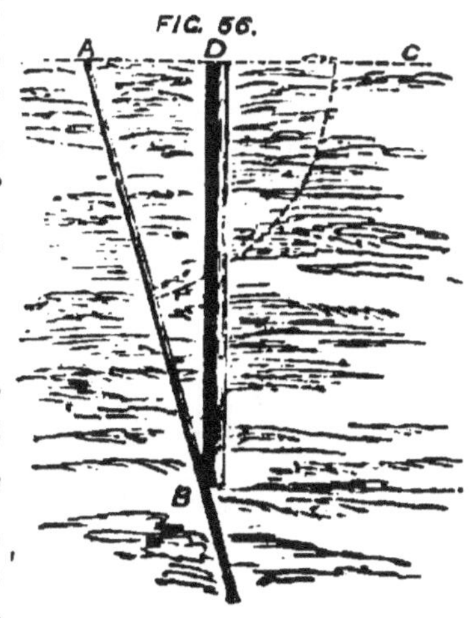

Ans. DB = 213·4 *yds.*

Example VI.—There is a vein of lead ore AB (see last fig.), which forms an angle CAB of 70° with the horizon, on which I wish to sink a shaft DB; I demand to know what distance AD the shaft must be set off from the vein at the surface, just to cut it at the depth of 141 yards?

The distance AD that the shaft must be set off at the surface from the vein AB, just to cut it at the depth of 141 yards, will be 51·3 yards.

Example VII.—I have to set a drift BC from the bottom of a pit AB, which is to be driven truly level; now I wish to know at what distance from the pit B it will cut the stratum of coal DE, which dips so as to form an angle aDE of 20° with the horizon, the drift BC being set from the bottom of the pit at B, 40 yards perpendicularly below the seam D, and driven in direction of the dip of the stratum?

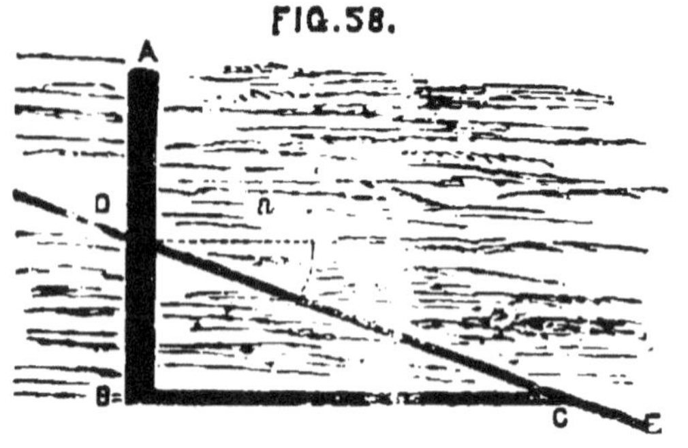

FIG. 58.

Ans. BC = 109·9 *yds.*

Example VIII.—I set off a drift at the side of a hill A, which was driven truly level, and cut a vein of lead ore at B, 100 yards distant from A, which vein I found to make an angle of 65° ABC with the horizon; now I wish to know what depth a shaft at A must be sunk just to cut the vein at C?

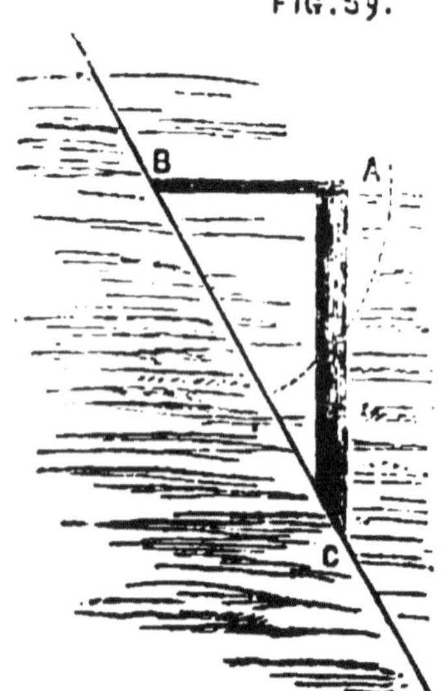

FIG. 59.

Ans. AC = 214·4 *yds.*

Example IX.—In the subterraneous survey ABCD, in the form of a trapezium, are two shafts F and E, joined by a fifth straight drift FE. Now this survey was made with a magnetic needle, which was afterwards found to be defective in its indications, on account of the presence of ferruginous substances both in the mine and on the surface; therefore, how is the work to be plotted, since the angles cannot be relied upon; and only the lengths of five drifts, and the segments of the drifts AB and CD made by the drift EF, are given; moreover, the tops of the shafts F and E range with the sun at 1½ P.M. on the 18th of October, 1860?

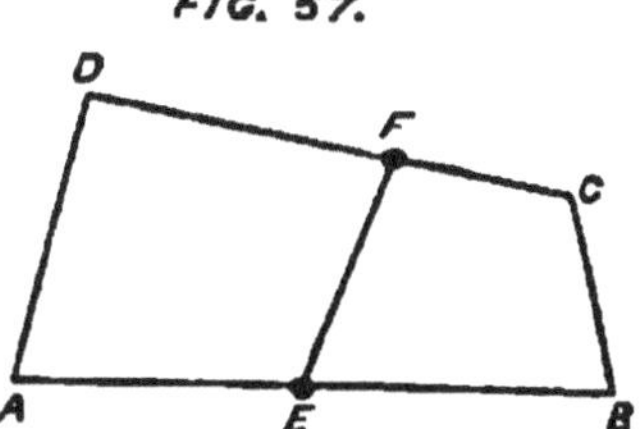

Note.—The solution of this question will require a knowledge of the application of algebra to geometry and of spherical trigonometry to astronomy.

EXAMPLE X.—There is a vein of lead ore AC, which forms an angle *aAe* of 80° with the horizon A*c*; now I have sunk a shaft AB on the vein at the surface, to the depth of 120 yards perpendicular; I desire to know what distance the bottom of the shaft B will be from the vein?

Ans. BC=21·1 *yds.*

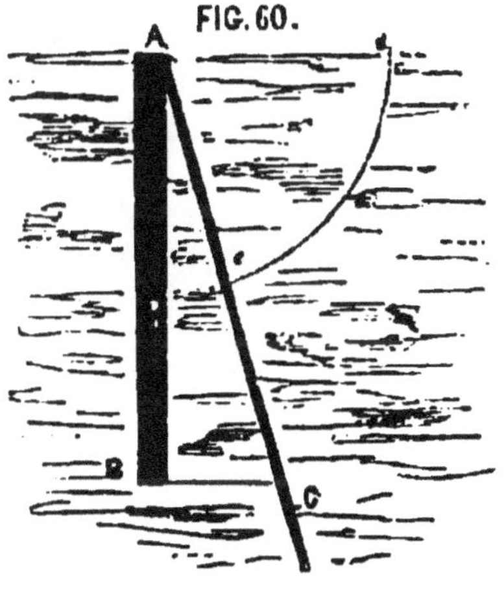

EXAMPLE XI.—I made a survey of a subterraneous excavation ABCDFG (see the following bearings and distances), commencing at the pit A; now I wish to know, on the surface, by one single bearing and distance, to be taken from the pit A, where I must sink a pit perpendicularly upon G, the extreme end of the excavation?

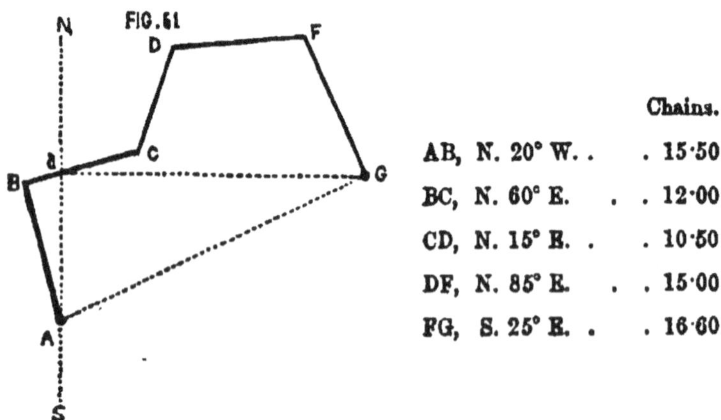

		Chains.
AB,	N. 20° W. .	. 15·50
BC,	N. 60° E. .	. . 12·00
CD,	N. 15° E. .	. 10·50
DF,	N. 85° E. .	. . 15·00
FG,	S. 25° E. .	. 16·60

The extreme point G of the excavation will have 16·98 chains of northing Aa, and 29·76 chains of easting aG, from A; therefore the line AG will be found to bear N 60° 17′ E with SN, the magnetic meridian of A, and its length will be 34·26 chains,—the bearing and distance required.

Note.—This Example ought also to be solved by taking the angles from the direction of the drift AB (the position of which is assumed to be fixed on the surface), by the method given in Art. 21.

EXAMPLE XII.—I have to drive an excavation from B towards A, which is to rise 1 inch in every 60 feet of its length; now I wish to put down an air-shaft P thereon, just 1600 yards from its mouth B; what will be the depth of the shaft from the surface to the sole of the drift or excavation, when the surface at P, the place where it has to be sunk, is 350 feet above the level of B, the mouth of the excavation?

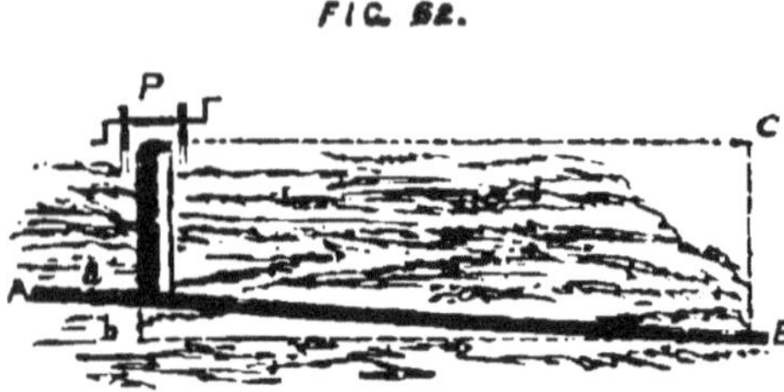

Ans. 343⅓ *feet.*

EXAMPLE XIII.—A headway is driven into a coal-stratum

from the foot of a hill; the straight portions of the headway, commencing at A, are AB = 5·63 chains, BC = 5·18, CD = 3·80, DE = 4·71, and EF = 7·02; the angles are ABC = 121° 16′, BCD = 228° 5′, CDE = 164° 52′, and DEF = 168° 29′, all the angles being taken by the theodolite on the right side of the lines in the headway. A shaft is required to be sunk on the hill to the coal-stratum, at the levelled or horizontal distance of 37·50 chains from the entrance of the headway, from which the top of the proposed shaft bears 37°42′ to the right of the direction of the first straight portion of the headway AB. Now the coal-stratum rises uniformly at the rate of 1¾ inches in a chain; it is required to find the direction and length of the additional headway from F to the bottom of the proposed shaft, also its depth, the angle of elevation of the top of the shaft from the entrance A of the headway being 12° 8′.

PART III.

THIS part of the work treats of those subjects which are particularly necessary to be attended to, because of the great number of existing surveys, which have been made by the help of the magnetic needle; and the consequent necessity of attending to the magnetic variation at the different periods of time, at which those surveys were made; of showing the method of finding the true meridian, and of determining the variation of different magnetic needles at different times, and of the manner of reducing bearings taken with the magnetic meridian to those formed with the true meridian. Various other subjects interesting to miners are here discussed, concluding with a Traverse Table, and the method of estimating the produce of seams of coal.

AXIOMS AND OBSERVATIONS.

(48.) 1.—Two magnetic needles seldom have exactly the same variation.

2.—The magnetic variation not being stationary, the variation of the needle of all instruments depending thereon will change accordingly.

3.—If a subterraneous survey is made by one instrument, and plotted on the surface by another, the needles of each having different magnetic variation, the plotting will be erroneous if the bearings to be plotted are not previously reduced to bearings with that magnetic needle by which it is to be plotted.

4.—If a subterraneous survey is made by one instrument, and plotted on the surface by another, the needles of both having the same magnetic variation, the survey will be truly plotted.

5.—If a survey is plotted on the surface immediately after it has been taken under ground (by the same instrument), no material error can result.

6.—If a survey is plotted on the surface by the same instrument it was made with, but at some distant time after, the plotting will be erroneous, inasmuch as the magnetic variation has changed in the time between the survey being taken underground and its being plotted on the surface.

7.—All bearings of subterraneous excavations which are added, from time to time, on any plan kept for that purpose, must be reduced to bearings with the delineated meridian of that plan, previous to their being plotted thereon, otherwise the plotting will be erroneous.

8.—All surveys of subterraneous excavations which are recorded for future purposes must be recorded with the variation of the needle by which they have been taken, or otherwise they must be reduced to bearings with the true meridian, and so recorded, notifying the same.

9.—All the preceding axioms and observations will be unnecessary in new surveys, which are made without the use of the needle, and which are unconnected with old surveys.

Of the magnetic variation of the needle.

(49.) Since nearly all subterraneous surveys have been made by, or have reference to, the magnetic needle, each bearing (as shown in the first part of this work) is taken by the angle it makes with the magnetic meridian; and that magnetic meridian has been continually changing at the rate of about 9 minutes annually, for 230 years, from north towards the west, up to 1793; but its annual declination was afterwards not so great, for the north end of the needle was little more than 24 degrees westward of the true meridian of London in 1803: At Paris it was somewhat less, still continuing to increase or decrease at a slow rate; while in some parts of the world the north end of the needle was even eastward of the true meridian at the last-named date. As the magnetic meridian is always changing, it must necessarily follow, that the same line which formed an angle with it, of a certain magnitude, on any particular day, will not (we have strong reasons to suppose) form the same angle that day twelve months with the then magnetic meridian: Hence follows the great necessity of reducing every bearing to the angle it will form with the true or invariable meridian; the manner of doing it will be shown hereafter: Also the records of subterraneous surveys noted down for future purposes, where the surveyor has neglected to insert from what kind of meridian the bearings thereof are formed; by such neglect those records will not only cease to be of use, but will tend to mislead.

I shall insert a table, showing the different degrees of magnetic variation at different times from the year 1575 to 1858, which is nearly to the present time:

VARIATION AT LONDON.

Year.	Variation.		Year.	Variation.	
1576	11° 15′		1745	16° 53′	
1580	11° 11		1750	17° 54′	
1612	6° 10′	E.	1760	19° 12′	
1622	6° 0′		1765	20° 0′	
1633	4° 5′		1770	20° 35′	
1657	0° 0′		1775	21° 28′	
1666	1° 35′		1777	21° 57′	
1672	2° 30′		1779	22° 4′	
1683	4° 30′		1780	22° 26′	W.
1692	6° 0′		1786	23° 19′	
1700	8° 0′		1789	23° 36′	
1717	10° 42′	W.	1793	23° 51′	
1724	11° 45′		1797	24° 2′	
1730	13° 0′		1800	24° 6′	
1735	14° 16′		1803	24° 9′	
1740	15° 40′		1806	24° 15′	

VARIATION AT LONDON.—*Continued.*

Year.	Variation.		Year.	Variation.	
1809	24° 22′	W.	1835	23° 32′	W.
1812	24° 28′		1838	23° 19′	
1815	24° 35′		1841	23° 6′	
1818	24° 41′		1844	22° 52′	
1820	24° 32′		1847	22° 41′	
1823	24° 20′		1850	22° 30′	
1826	24° 8′		1853	22° 19′	
1829	23° 56′		1856	22° 8′	
1832	23° 44′		1858	22° 2′	

Note.—By the variation being east or west, is meant that the north end of the magnetic needle is on the east or west side of the true meridian; and where the variation is called east or west in the following part of this work, it is to be understood that the north end of the magnetic needle has east or west variation accordingly, except it is particularly mentioned to the contrary.

From the table it appears the magnetic needle had east variation in the year 1576; that is, its north end was 11° 15′ on the east side of the true meridian of London; and in 1657 the needle was in direction of the true meridian; and since that time it has been veering about to the west, until it has got upwards of 24° to the westward thereof. Besides this annual variation just mentioned, it has a daily variation.

I shall insert a table, showing the diurnal variation taken at different hours of the 27th day of June, 1759, by Mr. Canton.—(*Phil. Trans.*, vol. 51.)

	Hrs.	Min.	Declination west.	Degrees of Fahrenheit's thermom.
Morning	0	18	18° 2′	62°
	6	4	18° 58′	62°
	8	30	18° 55′	65°
	9	2	18° 54′	67°
	10	20	18° 57′	69°
	11	40	19° 4′	68½°
Afternoon	0	50	19° 9′	70°
	1	38	19° 8′	70°
	3	10	19° 8′	68°
	7	20	18° 59′	61°
	9	12	19° 6′	59°
	11	40	18° 51′	57½°

The mean variation of each month of the year.—

January . . .	7′ 8″	July	13′ 14″
February . . .	8′ 58″	August . . .	12′ 19″
March . . .	11′ 17″	September . . .	11′ 43″
April	12′ 26″	October . . .	10′ 36″
May	13′ 0″	November . . .	8′ 9″
June	13′ 21″	December . . .	6′ 58″

To find the true meridian.

(50.) I shall lay down an easy and comprehensive rule to find the true meridian, which is preparatory to the determining of the magnetic variation of the needle. It is well known that the sun, at 12 o'clock at noon, is due south in all northern latitudes; and if a pole is set up perpendicular to the horizon, its shadow at that hour will bear exactly north, or in direction of the true meridian;—also the shadow of the pole will be shortest at that precise time.

Let ABC be a board perfectly plain and clear of twistings, and of a triangular form, each side about 30 inches long, having a number of concentric circles *cde* about 1½ inch asunder, drawn on its surface from a centre *a*. Now let this board be placed horizontal by means of a spirit level, with its angular point C towards the south; and at *a*, the centre of the concentric circles, let there be fixed an upright pin about 10 inches long, exactly perpendicular to the board, and also perpendicular to the horizon. All this being done on a clear day, and before the sun arrives on the meridian of the place of observation, which I shall say about 11 o'clock, then observe carefully the first concentric circle that the end of the shadow of the pin fixed at *a* touches, which suppose to be at *f*, and there make a mark: Then observe again carefully when the end of the same shadow touches on the same concentric circle, which will be about 1 o'clock,—suppose it to be at *g*; there make another mark: Then with a pair of compasses divide the distance *fg*, and the point in the middle between which, suppose *h*, will be the direction of the shadow of the pin at 12 o'clock: Consequently *ah* is the direction of the true meridian. Then by placing an upright sight E, with a slit *kk* in it, on the table, the centre of which coinciding with the point *h*, the pin at *a* having an opening in it similar to *bb*, with a perpendicular hair in direction of the opening; and by looking through the sight E, together with the hair in the centre of the opening in the pin placed at *a*, the meridian may be extended to any distance S on the surface; in the direction of which line it will be proper to place two permanent marks, as represented by NS, whose distance may be from 100 to 300 yards, for the purpose of determining at all times the magnetic variation of the needle of the different instruments, made use of in surveying: Such a line every director of mines ought to have marked out in the situation of the mine he directs.

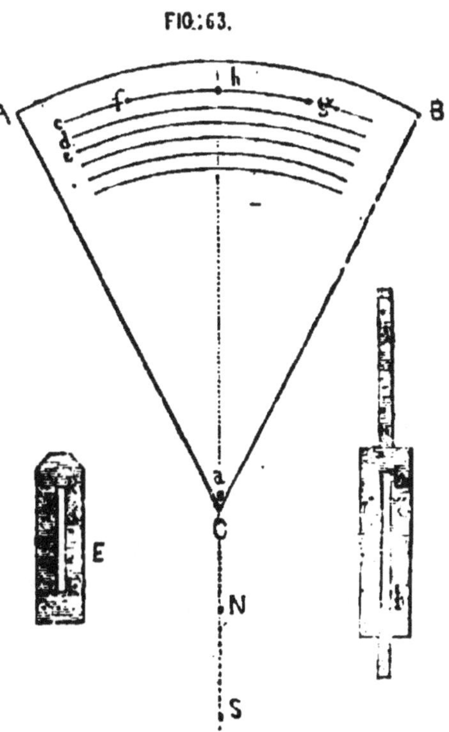

(50*a*.) If the student be acquainted with the application of spherical trigonometry to astronomy, he will find the following method of finding the true meridian to be greatly preferable to that just given. Let S represent the place of the sun's centre, P the north pole, and Z the zenith; these three points being the angles of a spherical triangle SPZ (the student can readily draw the figure for himself), in which SZ represents the co-altitude of the sun, when he comes into the direction of the required bearing of the drift in the mine; SP the sun's co-declination on the day of observation (which will be found in the Nautical Almanack for the year in which the observation is made); and PZ the co-latitude of the place of the mine (which is usually well known). From the given spherical triangle SPZ the angle Z may be readily found, which is the azimuth or bearing of the sun from the north at the time of observation, and also the bearing of the drift; whence also the true meridian may be readily deduced for the following purpose.

To determine the magnetic variation of the needle of any instrument.

(51.) Suppose N and S to be marks representing the true meridian, S the south and N the north; place the instrument (whose magnetic variation you would wish to know) at S, and turn the sights in direction of SN until N is seen through them; at the same time observe the bearing of the needle of the instrument, and whatever N is found to bear from due north, as much will the magnetic meridian differ from the true meridian. Suppose the north end of the needle to stand in direction of S*d*, then the true meridian SN will be to the east of the magnetic as much as the angle *d*SN, which suppose 23°; then SN will bear N 23° E with the magnetic meridian: Consequently the needle of the instrument may be said to have 23° of west variation, as the north end thereof is 23° to the west or left of the true meridian SN. Or if the north end of the needle stand in direction of S*e*, then the true meri-

FIG. 64.

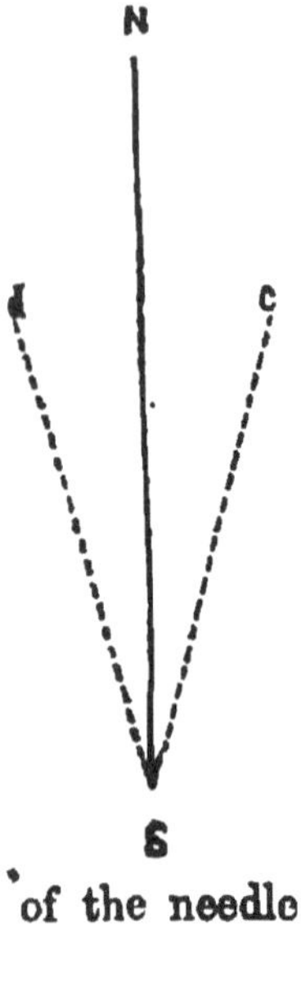

dian SN will be to the west of the magnetic as much as the angle *e*SN, which, if equal to 23°, then SN will bear N 23° W: Then the needle may be said to have 23° of east variation, the north end thereof being 23° to the east or right of the true meridian SN.

The manner of reducing bearings from a magnetic to a true meridian.

(52.) Let NS represent the true meridian, N the north and S the south, and *ns* a magnetic needle suspended on a centre *c*, representing the magnetic meridian, *n* the north and *s* the south; then the arch *na* will be the variation of the magnetic meridian from the true meridian, which may be called west variation, the north end of the needle being to the west side of the true meridian: And if the angle *nca* is equal to 23°, then the needle will have 23° of west variation, and the south end *s* will have 23° of east variation; for *s* will be to the east of the true south meridian line as much as the north end *n* is to the west of the true north meridian line.—(See theorem 8.)

FIG. 65.

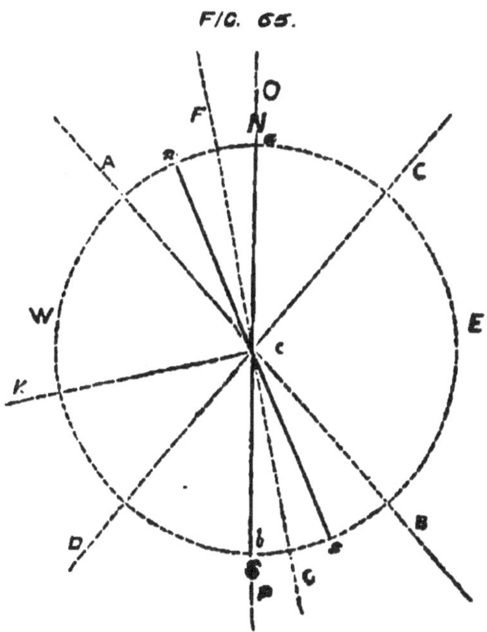

1st.—Suppose the circle WE to represent a circumferentor, and that the bearing of the object O with the true meridian is required; if *ns* is the needle representing the magnetic meridian, and the object O is found to form an angle *nca* with it of 23°, which (from the manner of determining bearings, Art. 2) may be called N 23° E, and, as before, the magnetic variation of the needle being 23° to the west of the true meridian, then 23° — 23° = 0; therefore the bearing of O with the true meridian S*c*N will be due north, for the needle ought to have stood in direction *ab*.

2d.—Suppose again the bearing of the object A with the true meridian is required; the bearing of A with the magnetic meridian will be equal to the angle *nc*A, which call N 10° W; but as the magnetic meridian has 23° of west variation, the bearing of A with the true meridian will be N 23° + 10° = 33° W; for angle *ac*A is equal to 33°, which is the angle that *c*A makes with the true meridian S*c*N.

3d.—Suppose again the bearing of the object C with the true meridian is required; the bearing of C with the magnetic meridian will be equal to the angle *nc*C, which call N 53° E; but the variation of the needle being 23° to the west of true north, and ought to have stood in the direction of *ab*, consequently the bearing of C from *c* with the true meridian will be N 53° — 23° = 30° E; for angle *ac*C is equal to 30°, which is the angle that the line *c*C makes with the true meridian line S*c*N.

4th.—Suppose the bearing of the object D with the true meridian is required; the bearing of D with the south magnetic meridian will be equal to the angle *sc*D, which call S 56° W; but the south end of the needle having 23° of east variation, and ought to have stood in direction of *ab* the true meridian, consequently the bearing of D from *c* with the true meridian will be S 56° – 23° = 33° W; for angle *bc*D is equal to 33°, which is the angle that the line *c*D makes with the true meridian line N*c*S.

5th.—Suppose again the bearing of the object B with the true meridian is required; the bearing of B with the magnetic meridian will be equal to the angle *sc*B, which call S 15° E; but the south end of the needle having 23° of east variation, consequently the true bearing of B will be S 15° + 23° = 38° E; for angle *bc*B is equal to 38°, which is the angle that the line *c*B makes with the true meridian line N*c*S.

6th.—Suppose again the bearing of the object F with the true meridian is required; the bearing of F with the magnetic meridian will be equal to the angle *nc*F, which call N 13° E; but the magnetic meridian has 28° of west variation, consequently the bearing of F with the true meridian will be N 23° − 13° = 10° W; for angle *ac*F is equal to 10°, which is the angle that the line *c*F makes to the left with the true meridian S*c*N.

7th.—Suppose again the bearing of the object G with the true meridian is required; the bearing of G with the magnetic meridian will be equal to the angle *sc*G, which call S 13° W; but the south magnetic meridian has 23° of east variation, consequently the bearing of G with the true meridian will be S 23° − 13° = 10° E; for angle *bc*G is equal to 10°, which is the angle that the line *c*G makes to the right with the true meridian N*c*S.

8th.—Suppose again the bearing of the object K with the true meridian is required; the bearing of K with the magnetic meridian will be equal to the angle *nc*K, which call N 80° W; the magnetic meridian having 23° of west variation, the angle that *c*K will make with the true north meridian *c*N will be 80° + 23° = 103°, *ac*K; but as it exceeds 90°, therefore 180° − 103° = 77°, angle *bc*K; then the bearing of K with the true meridian will be S 77° W; for angle *bc*K is equal to 77°, which is the angle that the line *c*K makes with the true south meridian line N*c*S.

N.B.—The true bearing of any object is nothing more than the angle that the object makes with the true meridian, instead of the angle it forms with the magnetic meridian; therefore, by the several cases of Art. 52, the method of solving the following examples will be readily seen:

EXAMPLE I.—If the following bearings, N 20° W, N 60°

E, N 70° W, and N 13 E are taken by an instrument whose magnetic needle has 23° west variation, what will be their bearings with the true meridian?

The first bearing N 20° W will form a bearing of N 20° + 23° = 43° W with the true meridian.

The second bearing, N 60° E, will form a bearing of N 60 − 23° = 37° E with the true meridian.

The third bearing, N 70° W, will form a bearing of 180° − 70° + 23° = 87°, which will be S 87° W with the true meridian.

The fourth bearing, N 13° E, will form a bearing of N 23° − 13° = 10° W with the true meridian.

With the magnetic meridian.	*With the true meridian.*
Thus, N. 20° W.	N. 43° W.
N. 60° E.	N. 37° E.
N. 70° W.	S. 87° W.
N. 13° E.	N. 10° W.

EXAMPLE II.—If the following bearings are taken by a meridian having 23° of west variation—S 10° W, N 10° E, N 50° E, and N 20° W—what will be their bearings with the true meridian?

With the magnetic meridian.	*With the true meridian.*
S. 10° W.	S. 13° E.
N. 10° E.	N. 13° W.
N. 50° E.	N. 27° E.
N. 20° W.	N. 43° W.

EXAMPLE III.—If the following bearings are taken by a meridian having 10° of west variation—N 50° W, N 70° E, S 5° E, and S 60° W—what will be their bearings with the true meridian?

With the magnetic meridian.	*With the true meridian.*
N. 50° W.	N. 60° W.
N. 70° E.	N. 60° E.
S. 5° E.	S. 15° E.
S. 60° W.	S. 50° W.

EXAMPLE IV.—If the bearings in the last example be taken by a meridian having 6° of east variation, what will be their bearings with the true meridian?

With the magnetic meridian.	*With the true meridian.*
N. 50° W.	N. 44° W.
N. 70° E.	N. 76° E.
S. 5° E.	S. 1° W.
S. 60° W.	S. 66° W.

The manner of reducing a bearing from one magnetic meridian to its bearing with any other magnetic meridian of different variation.

(53.) 1st.—Suppose the bearing of the object P from C is taken by a circumferentor whose needle has 10° of west variation *n's'*, which bearing is to be reduced to the bearing it will form with another magnetic meridian *ns*, having 23° of west variation: Let NS represent the true

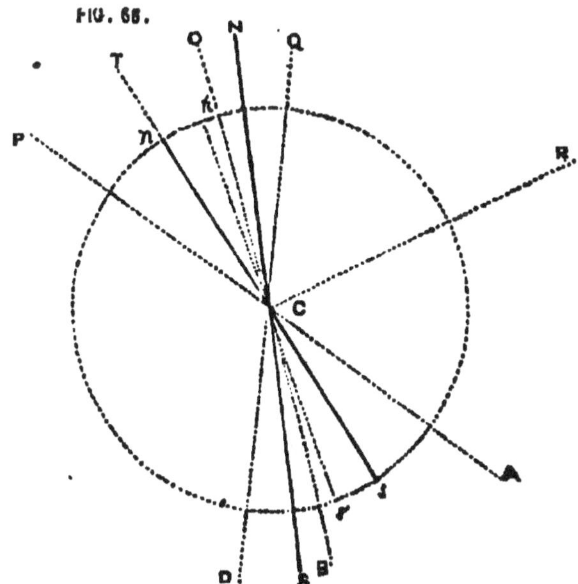

meridian, and the bearing of CP therewith (from the manner of reducing bearings, &c., Art. 52) equal to the angle PCN 45°, or N 45° W; also the magnetic meridian to which the bearing PC is to be reduced equal to the angle *n*CN 23°, or having 23° of west variation; the object P and the magnetic variation of the meridian to which its bearing is to be reduced are both on the *west side* of the true meridian NS; therefore ∠ PCN 45° − ∠ *n*CN 23° = ∠ PC*n* 22°; and as the angle PCN exceeds the angle *n*CN, the object P from C must bear N 22° W with the magnetic meridian *ns*.

2d.—Suppose the bearing of the object O from C is taken by an instrument whose needle has 10° of west variation *n′s′*, which is to be reduced to the bearing it will form with another magnetic meridian *ns*, having 23° of west variation: Let the bearing of CO with the true meridian be found equal to the angle OCN 8°, or N 8° W; and the magnetic meridian to which the bearing CO is to be reduced equal to the angle *n*CN 23°, or having 23° of west variation; the object O and the magnetic variation of the meridian to which its bearing is to be reduced are both on the *west side* of the true meridian NS; therefore ∠ *n*CN 23° − ∠ OCN 8° = ∠ OC*n* 15°; and as the angle OCN is less than the anlge *n*CN, the bearing of O from C will be N 15° E with the magnetic meridian *ns*.

3d.—Suppose the bearing of the object T from C is taken by a magnetic needle having 10° of west variation *n′s′*, which bearing is to be reduced to the bearing it will form with another magnetic meridian *ns*, having 23° of west variation: Let the bearing of TC with the true meridian be found equal to the angle TCN 23°, or N 23° W; and the magnetic meridian to which the bearing TC is to be reduced equal to the angle *n*CN 23°, or having 23° of west variation; then ∠ TCN 23° − ∠ *n*CN 23° = 0°; therefore the bearing of T from C will be in the direction of the magnetic meridian *ns*, or due north.

4th.—Suppose the bearing of the object Q from C is taken by a magnetic needle having 10° of west variation *n′s′*, which bearing is to be reduced to the bearing it will form with another magnetic meridian, *ns*, having 23° of west variation: Let the bearing QC with the true meridian NS be found equal to the angle QCN 15°, or N 15° E; and the magnetic meridian to which the bearing QC is to be reduced equal to the angle *n*CN 23°, or having 23° of west variation; now the object Q and the magnetic variation of the meridian to which its bearing is to be reduced are on *contrary sides* of the true meridian NS; therefore ∠ QCN 15° + ∠ *n*CN 23° = ∠ QC*n* 38°; and also the bearing of Q will be on the *contrary side* of that magnetic meridian *ns* that its variation is on; and as *ns* has west variation, therefore the bearing of Q from C will be N 38° E with the meridian *ns*.

5th.—Suppose the bearing of the object A from C is taken by the meridian *n′s′*, having 10° of west variation, which is to be reduced to the bearing it will form with another magnetic meridian *ns*, having 23° of west variation: Let the bearing of CA with the true meridian NS be found equal to the angle ACS 45°, or S 45° E; and the south magnetic meridian to which the bearing AC is to be reduced equal to the angle *s*CS 23°, or having 23° of east variation (see theorem 3, Art. 48); the bearing of the object A and the magnetic variation of the meridian to which its bearing is to be reduced are both on the *east side* of the true meridian; therefore ∠ ACS 45° − ∠ *s*CS 23° = ∠ AC*s* 22°; the angle ACS exceeding the angle *s*CS, the bearing of A with the magnetic meridian *ns* will be S 22° E.

6th.—Suppose the bearing of the object B from C is taken by a needle *n′s′*, having 10° of west variation, which is to be reduced to its bearing with another magnetic meridian *ns*, having 23° of west variation: Let the bearing of CB with the true meridian NS be found equal to the angle BCS 8°, or S 8° E; and the south magnetic meridian to which the bearing BC is to be reduced equal to the angle *s*CS 23°, or having 23° of east variation (see theorem 3,

Art. 48); the bearing of the object B and the magnetic variation of the meridian to which its bearing is to be reduced are both on the *east side* of the true meridian; therefore ∠ *s*CS 23°—∠ BCS 8° = ∠ BC*s* 15°; and as the angle BCS is less than the angle *s*CS, the bearing of B from C will be S 15° W with the magnetic meridian *ns*.

7th.—Suppose the bearing of the object D from C is taken by a needle *n's'*, having 10° of west variation, which is to be reduced to its bearing with another magnetic meridian *ns*, having 23° of west variation: Let the bearing of CD with the true meridian NS be found equal to the angle DCS 15°, or S 15° W; and also the south magnetic meridian to which the bearing DC is to be reduced equal to the angle *s*CS 23°, or having 23° of east variation (see theorem 3, Art. 48); and as the bearing of the object D and the magnetic variation of the meridian to which its bearing is to be reduced are on *contrary sides* of the true meridian NS, therefore ∠ DCS 15° + *s*CS 23° = ∠ DC*s* 38°; and also the bearing of D will be on the *contrary side* of the magnetic meridian *ns* that its variation is on; and as the south meridian *ns* has east variation, therefore the bearing of D from C will be S 38° W.

8th.—Suppose the bearing of the object R from C is taken by the meridian *n's'*, having 10° of west variation, which is to be reduced to the bearing it will form with another magnetic meridian *ns*, having 23° of west variation: Let the bearing RC with the true meridian be found equal to the angle RCN 77°, or N 77° E; and the magnetic meridian to which the bearing RC is to be reduced equal to the angle *n*CN 23°, or having 23° of west variation; the bearing of the object R and the magnetic variation of the meridian to which it is to be reduced are on *contrary sides* of the true meridian NS; therefore ∠ RCN 77° + ∠ *n*CN 23° = ∠ RC*n* 100°; but as the angle that the object R makes with the north magnetic meridian *ns* exceeds 90°, its bearing in that case must be with the *south or contrary* meridian; then 180° − 100° = 80° / RCs; consequently the bearing of the object R with the magnetic meridian ns will be S 80° E.

Note.—From the several cases of Art. 53, the student will have no difficulty in solving the following examples, with respect to two different magnetic variations.

EXAMPLE I.—If the following bearings are taken by a meridian having 10° of west variation,—N 50° W, N 70° E, S 5° E, and S 80° E; what will be the bearing of each with a meridian having 23° of west variation?

With a meridian of 10° *of variation.*	*With a meridian of* 23° *of variation.*
N. 50° W.	N. 37° W.
N. 70° E.	N. 83° E.
S. 5° E.	S. 8° W.
S. 80° E.	S. 67° E.

EXAMPLE II.—If the following bearings are taken by a meridian having 10° of east variation,—S 60° W, S 10 E, N 80° E, and N 10° W; what will be the bearing of each with a meridian having 20° of west variation?

With a meridian of 10° *of east variation.*	*With a meridian of* 20° *of west variation.*
S. 60° W.	Due west.
S. 10° E.	S. 20° W.
N. 80° E.	S. 70° E.
N. 10° W.	N. 20° E.

EXAMPLE III.—The following bearings are taken by a meridian having 20° of west variation,—S 60° W, N 5° W, N 30° W, and N 50° E; what bearing will each form with a meridian having 10° of east variation?

With a meridian of 20° *of west variation.*	*With a meridian of* 10° *of east variation.*
S. 60° W.	S. 30° W.
N. 5° W.	N. 35° W.
N. 30° W.	N. 60° W.
N. 50° E.	N. 20° E.

EXAMPLE IV.—If the following bearings are taken by the true meridian,—S 60° W, N 5° W, N 30° W, and N 50 E; what bearing will each form with a meridian having 23° of west variation?

With the true meridian.	*With a meridian of 23° of west variation.*
S. 60° W.	S. 83° W.
N. 5° W.	N. 18° E.
N. 30° W.	N. 7° W.
N. 50° E.	N. 73° E.

EXAMPLE V.—I have to plot a survey on the surface of the following bearings and distances,—N 25° W 5 chains, N 63° W 10 chains, N 20° E 3 chains, N 70° E 6 chains, and S 84° E 9 chains, which has been taken by a circumferentor having 20° of west variation; now I find the circumferentor by which I have to plot the same has 23° of west variation, I demand to know the bearings under which the survey must be plotted, so that the same may be accurately done?

With a meridian of 20° of west variation.	Chains.	*The bearings under which the survey must be plotted to be accurately done, by a needle having 23° of west variation.*	Chains.
N. 25° W.	5	N. 22° W.	5
N. 63° W.	10	N. 60° W.	10
N. 20° E.	3	N. 23° E.	3
N. 70° E.	6	N. 73° E.	6
S. 84° E.	9	S. 81° E.	9

EXAMPLE VI.—In a subterraneous survey of the following bearings and distances, viz. N 20° W 10 chains, N 60° W 3 chains, S 12° W 5 chains, N 87° W 4 chains, and S 15° E 7 chains, surveyed by an instrument having 22° of west variation, which is to be plotted on a plan whose meridian has 12° of west variation, I wish to know under what bearing each must be plotted on the plan, so that it may be accurately done?

The bearings by a meridian having 22° of west variation.	Chains.	*The bearings with the plan's meridian having 12° of west variation.*	Chains.
N. 20° W.	10	N. 30° W.	10
N. 60° W.	3	N. 70° W.	3
S. 12° W.	5	S. 2° W.	5
N. 87° W.	4	S. 83° W.	4
S. 15° E.	7	S. 25° E.	7

To find what kind of a meridian a plan has been constructed by.

(54.) Where subterraneous excavations are to be added to some previously delineated on a plan, it will be necessary, first of all, to find what kind of meridian the plan has been constructed by, in order that the bearings to be plotted may previously be reduced thereto (see theorem 4, Art. 48).

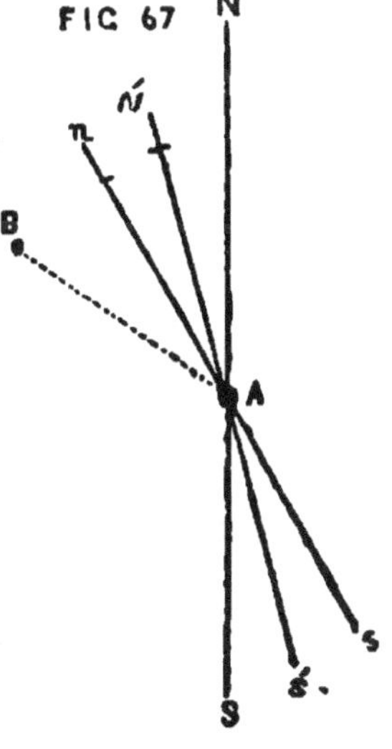

1. Suppose $N'S'$ to be the meridian of a plan whose magnetic variation is required to be known; let the bearing of the pit B from the pit A be taken on the plan with the meridian thereon, equal to the angle BAN' 40°, or N 40° W; and let the bearing of the same two pits be taken on the surface by a circumferentor placed at A, whose needle is known to have 23° of west variation ns, and found to form an angle BAn = 27°, or N 27° W; then, if $N'S'$ represent the true meridian, the line AB will form an angle therewith of 27° + 23° = 50° BAN, or N 50° W: From ∠ BAN 50 ∠

BAN' 40°, leaves ∠ N'AN = 10°, which is the angle that the plan's meridian makes with the true meridian; and as the angle BAN', which is the bearing of the object with the plan's meridian, *is to the left* thereof, and less than the ∠ BAN, which is the bearing of the same object, as taken by the circumferentor on the surface, with the true meridian, and *to the left* thereof also, it follows that ∠ N'AN, the variation of the plan's meridian, must be *to the left* of the true meridian; therefore $S'N'$ must have 10° of west variation.

2. Suppose $N'S'$ to be the meridian of a plan whose magnetic variation is required to be known; let the bearing of the pit B from A be taken on the plan with the meridian thereon, equal to the angle BAN' 60°, or N 60° W; and let the bearing of the same two pits be taken on the surface by a circumferentor placed at A, whose needle is known to have 23° of west variation *ns*, and found to form an angle BAn = 27°, or N 27° W; then if NS represent the true meridian, the line AB will form an angle therewith of 27° + 23° = 50° BAN, or N 50° W: Then from ∠ BAN' 60° − ∠ BAN 50°, leaves ∠ N'AN = 10°, the variation of the plan's meridian; but as the ∠ BAN', which the bearing of the object makes *to the left* with the plan's meridian, is greater than the ∠ BAN, which is the angle that the same object, as taken by the circumferentor on the surface, makes *to the left* with the true meridian, the ∠ N'AN must be *to the right* of the true meridian; therefore $S'N'$ must have 10° of west variation.

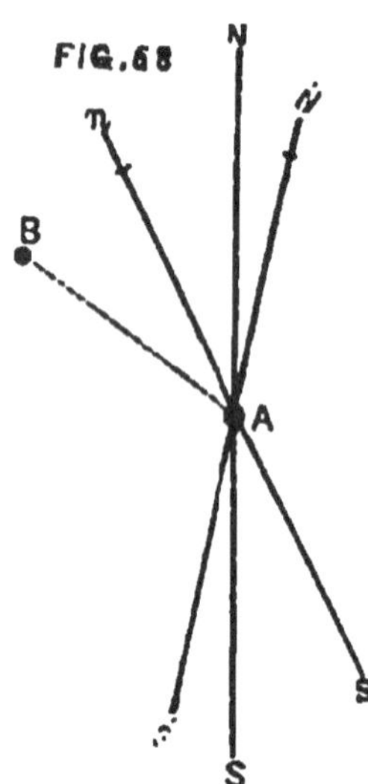

3. Suppose $N'S'$ to be the meridian of a plan whose magnetic variation is required to be known; let the bearing of the pit B from the pit A be taken on the plan with the

meridian thereon, equal to the angle BA*N′* 5°, or N 5° E; and let the bearing of the same two pits be taken on the surface by a circumferentor placed at A, whose needle has 23° of west variation *ns*, be found to form an angle BA*n* = 18°, or N 18° E; then if NS represent the true meridian, the line AB will form an angle therewith of 23° − 18° = 5° ∠ BAN, or N 5° W: Then ∠ BA*N′* 5° + ∠ BAN 5° = ∠ *N′*AN 10°, the variation of the plan's meridian; and as AB bears on different sides of the two meridians *N′S′* and NS, and ∠ BAN being *to the left* of the true meridian NS, ∠ NA*N′* must be *to the left* thereof also; consequently the plan's meridian *N′S′* must have 10° of west variation.

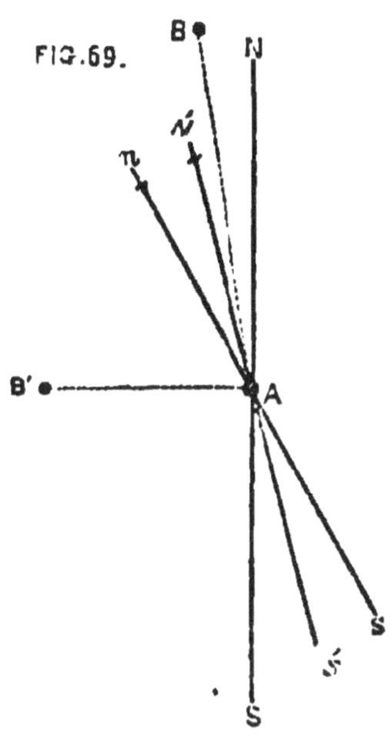

4. Suppose *N′S′* (see last fig.), is the meridian of a plan whose magnetic variation is required to be known; let the bearing of the pit B′ from the pit A be taken on the plan with its meridian, equal to the angle *N′*AB′ 83°, or N 83° W; and let the bearing of the same two pits be taken on the surface by a circumferentor placed at A, whose needle has 23° of west variation *ns*, be found to form an angle *n*AB′ = 70°, or N 70° W; then if NS represent the true meridian, the line AB′ will form an angle therewith of 87° ∠ B′AS S 87° W (see Art. 52): Now ∠ *N′*AB′ 83° + ∠ B′AS 87° = ∠ *N′*AS 170°, then 180° − 170° = 10° ∠ NA*N′*, the variation of the plan's meridian; and as ∠ NA*N′* 10° is what ∠ SA*N′* falls short of 180°, reckoning from the south meridian S, therefore it must be *to the left or west* of the north meridian N; consequently the plan's meridian *N′S′* must have 10° of west variation.

5. Suppose $N'S'$ to be the meridian of a plan whose magnetic variation is required; let the bearing of the pit B from the pit A be taken on the plan with its meridian thereon, equal to the angle N'AB 45°, or N 45° W; and let the bearing of the same two objects, taken on the surface by an instrument placed at A, whose needle has 23° of west variation *ns*, be found to be equal to the same angle *n*AB 45°, or N 45° W, as before; then if NS represent the true meridian, the line AB will form an angle therewith of 45° + 23° = 68° ∠ NAB, or N 68° W: Then ∠ NAB 68° − ∠ N'AB 45° = ∠ NAN' 23°, the variation of the plan's meridian; but as ∠ N'AB is *to the left* of the plan's meridian, and is less than ∠ NAB, the ∠ NAN' must be *to the left* of the true meridian SN; therefore the plan's meridian, will have 23° of west variation.—When the bearing of two objects, taken on a plan by its delineated meridian, agrees with the bearing of the same two objects taken on the surface by an instrument, the variation of the plan's meridian will be the same as the magnetic variation of the needle of that instrument.

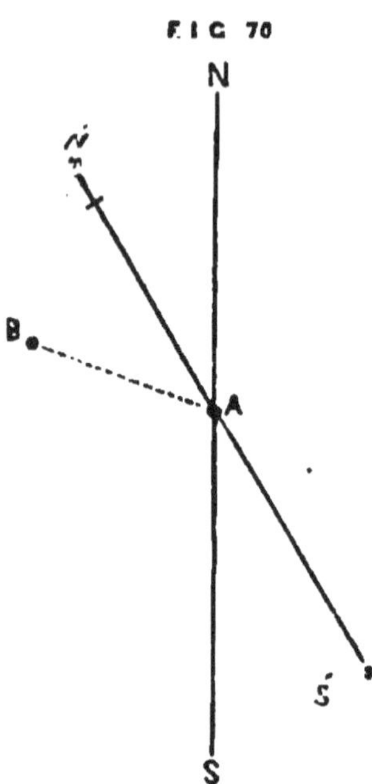

6. Suppose $N'S'$ to be the meridian of a plan whose magnetic variation is required to be known; let the bearing of the pit B from that of A be taken on the plan by its meridian thereon, equal to the anglo NAB 68°, or N 68° W: and let the bearing of the same two objects be taken by an instrument on the surface placed at A, whose needle has 23° of west variation *ns*, equal to the angle *n*AB 45°, or N 45° W; then the object will form an angle with the true

meridian of 45° + 23° = 68°, or N 68° W: Now, as the bearing of the two objects on the plan with its meridian, agrees with the bearing of the same two objects taken on the surface when reduced to the true meridian, therefore the plan's meridian must be the true meridian.

From the several cases in the last Article, where six examples are solved, the method of solving the following unsolved examples will be readily seen.

Example I.—I wish to know the variation of a plan's meridian, when the bearing of two objects thereon with its meridian is N 30° W, and the bearing of the same two objects with each other on the surface is found, by an instrument whose needle has 20° of west variation, to be N 19° W?

The objects on the surface will form a bearing with each other of N 39° W by the true meridian.

Then 39° — 30° = 9°; therefore the plan's meridian has 9° of west variation.

Example II.—I wish to know the variation of a plan's meridian, when the bearing of two objects thereon with its meridian is N 16° E, and the bearing of the same two objects with each other on the surface is found, by an instrument whose needle has 23° of west variation, to be N 10° E?

The objects on the surface will form a bearing with each other of N 13° W by the true meridian.

The 16° + 13° = 29°; therefore the plan's meridian has 29° of west variation.

Example III.—I have a plan which I wish to know by what kind of meridian it has been delineated: Now the bearing of two objects thereon with each other by its meridian is found to be N 80° W, and the bearing of the same two objects, taken on the surface by an instrument whose needle has 21° of west variation, is N 74° W?

The bearing of the two objects on the surface with the true meridian will be S 85° W.

Then $180^\circ - \overline{80^\circ + 85^\circ} = 15^\circ$; therefore the plan has been delineated by a meridian having 15° of west variation.

EXAMPLE IV.—I wish to know the variation of a plan's meridian, when the bearing of two objects taken thereon by its meridian is found to be N 40° E, and the bearing of the same two objects, taken on the surface by an instrument whose needle has 20° of west variation, is also N 40 E?

Then the meridian of the plan will have the same magnetic variation as the needle by which the bearing of the objects was taken on the surface; therefore the plan's meridian will have 20° of west variation.

EXAMPLE V.—I wish to know by what kind of meridian a plan has been constructed, when two objects thereon by its meridian form a bearing with each other of N 32° W, and the bearing of the same two objects, as taken on the surface by an instrument whose needle has 22° of west variation, forms a bearing with each other of N 10° W?

The two objects on the surface will form a bearing with each other of N 32° W by the true meridian.

Then the meridian of the plan will be the true meridian.

EXAMPLE VI.—I wish to know the variation of a plan's meridian, when the bearing of two objects thereon with its meridian is S 16° W, and the bearing of the same two objects with each other on the surface, taken by an instrument whose needle has 23° of west variation, is found to be S 10° W?

The plan's meridian will have 29° of west variation.

EXAMPLE VII.—I wish to know the variation of a plan's meridian, when the bearing of two objects thereon with its meridian is S 40° W, and the bearing of the same two objects with each other on the surface, taken by an instrument whose needle has 20° of west variation, is found to be S 23° W?

The plan's meridian will have 6° of east variation.

EXAMPLE VIII.—I wish to know the variation of a plan's meridian, when the bearing of two objects thereon with its meridian is N 65° W, and the bearing of the same two objects with each other on the surface, taken by an instrument whose needle has 23° of west variation is found to be N 20° W?

The plan's meridian will have 22° of east variation.

EXAMPLE IX.—I have a plan of a colliery workings, on which I took the bearing of two pits with each other by its meridian, which was N 5° W; I also took the bearing of the same two pits on the surface by an instrument whose needle had 23° of west variation, which was N 5° E; now I wish to know the variation of the plan's meridian by which it has been delineated?

The plan's meridian will have 13° of west variation.

EXAMPLE X.—I wish to know by what kind of meridian a plan of a colliery working has been constructed, when the bearing of two pits thereon with each other by its delineated meridian is found to be N 5° E, and the bearing of the same two pits on the surface with the true meridian is found to be N 14° W?

The plan has been constructed by a meridian having 19° of west variation.

How to plan surveys, and also the manner of determining an error arising in plotting, through inattention to the magnetic variation of the needle.

(55.) It has been shown, in Art. 49, that the magnetic meridian is always changing; therefore the bearings of the same objects, taken by such a meridian at different times, must also vary from each other, except reduced to bearings with the true meridian.

Let NS represent the meridian of a plan, which is also supposed to be the true meridian; and if a subterraneous excavation is to be plotted thereon from the pit A, which excavation is found to form a bearing of N 10° W 10 chains by an instrument whose needle had 20° of west variation;

now if the excavation N 10° W 10 chains is plotted on the plan by its meridian NS, which is the true meridian, it will be represented by AB; but the bearing being taken by a needle having 20° of west variation, therefore (according to the manner of reducing bearings from one magnetic meridian to their bearings with any other, Art. 53) it should form a bearing of N 30° W with the meridian NS, as represented by A*b*; then A*b* will be the true direction of the excavation from the pit A, and *b*B will be the magnitude of the error (see theorem 8, Art. 48): Or, instead of reducing the excavation to its bearing with the true meridian NS, it will be equally as true if *ns* is drawn on the plan, and made to represent the magnetic meridian of the needle by which the bearing was taken, with which A*b* will form a bearing of N 10° W.

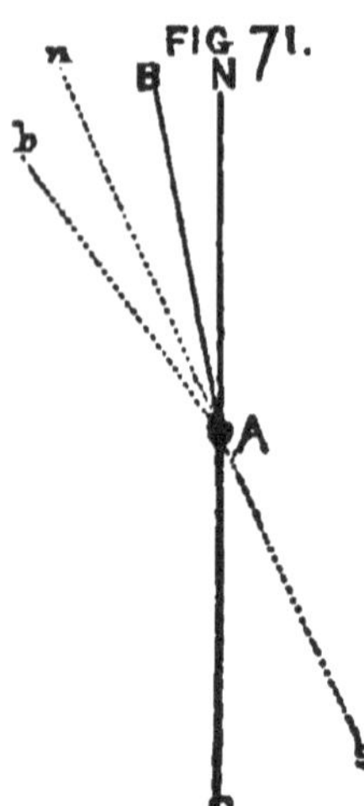

I shall insert a few examples, illustrative of the error arising from plotting a subterranous survey on a plan without attending to the variation of the magnetic meridian, and also how its magnitude can be ascertained.

Example I.—The following is a subterraneous survey, commencing at a pit called the B pit, N 30° W 6 chains, N 70° E 10 chains, N 30° E 5 chains, and N 25° W 8 chains, which was surveyed by an instrument whose needle had 24° of west variation; under what bearings must the survey be plotted on a plan whose delineated meridian has 15° of west variation?

Reduce the bearings, as taken by a meridian having 24° of west variation; to bearings with a meridian having 15° of west variation: Thus,—

Bearings with a meridian of 24° of west variation.	Chains.	*Bearings with a meridian of 15° of west variation.*	Chains.
N. 30° W. . .	6	N. 39° W. . .	6
N. 70° E. . .	10	N. 61° E. . .	10
N. 30° E. . .	5	N. 21° E. . .	5
N. 25° W. . .	8	N. 34° W. . .	8

The survey must be plotted under bearings with a magnetic meridian having 15° of west variation, as above, commencing at the B pit.

EXAMPLE II.—If the following subterraneous survey, N 9°.W 8 chains, N. 30° E 7 chains, and N 21° W 8 chains, is made by an instrument whose needle has 23° of west variation, and plotted on a plan by a meridian having 5° of west magnetic variation, without being reduced thereto,—what will be the magnitude of the error resulting by such neglect?

FIG. 78.

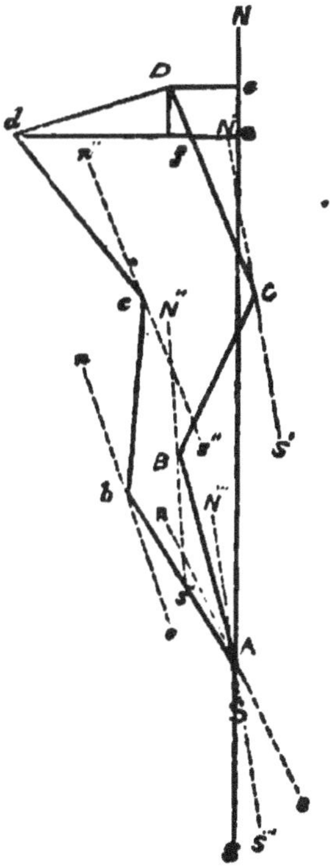

Suppose A, the point of commencement of the survey on the plan, and let the meridian of the plan here presented be $N'''S'''$, having 5° of west variation with the true meridian NS; then the first bearing, N 9° W 8 chains, will be represented by AB, — the second, N 30° E 7 chains, by BC,—and the third bearing, N 21° W 8 chains, by CD; then ABCD will represent the survey plotted without attending to the magnetic variation: But as the survey was made by an instrument whose needle had 23° of west variation, therefore each bearing, when truly plotted, must be set off from a meridian of that variation, which let *ns* represent;

then N 9° W 8 chains will be represented by A*b*, N 30° E 7 chains by *bc*, and N 21° W 8 chains by *cd*; then A*bcd* will represent the survey truly plotted, and *d*D will be the magnitude of the error.

Or the survey may be plotted by reducing the bearings, as taken by a meridian of 23° of west variation, to bearings, with a meridian of 5° of variation, as represented by *N′S′*, and plotted from it accordingly,— which will exactly coincide with A*bcd*, as before.

To discover, by calculation, the magnitude of the error, reduce the bearings of the survey, as taken by a magnetic meridian having 23° of west variation, to bearings with the true meridian,—and also the same bearings, as if taken by a meridian having 5° of west variation, to bearings with the true meridian; then determine the northing and easting of D from *d*: Thus,—

With a meridian of 23° of west variation.	With the true meridian.	With a meridian of 5° of west variation.	With the true meridian.
Chns.	Chns.	Chns.	Chns.
N. 9° W. 8	N. 32° W. 8	N. 9° W. 8	N. 14° W. 8
N. 30° E. 7	N. 7° E. 7	N. 30° E. 7	N. 25° E. 7
N. 21° W. 8	N. 44° W. 8	N. 21° W. 8	N. 26° W. 8

	Northing.	Southing.	Easting.	Westing.	
Chns.	Chains.	Chains.	Chains.	Chains.	
N. 32° W. 8	6·78	...	...	4·23	
N. 7° E. 7	6·94	...	0·85	...	
N. 44° W. 8	5·75	...	..	5·55	
	19·47	A*a*		9·78	
				0·85	
				8·93	*ad*

	Northing.	Southing.	Easting.	Westing.	
Chns.	Chains.	Chains.	Chains.	Chains.	
N. 14° W. 8	7·76	...	...	1·93	
N. 25° E. 7	6·34	...	2·95	...	
N. 26° W. 8	7·19	...	...	3·50	
	21·29	Ae		5·43	
				2·95	
				2·48	eD or af

ad 8·93 chains — *af* 2·48 chains = *fd* 6·45 chains.
A*e* 21·29 chains — A*a* 19·47 chains = *ae* or *f*D 1·82 chains.

Then, as *fd* 6·45 . . .	·8095595
Is to radius	10·0000000
So is *f*D 1·82 . . .	·2600714
To tang. ∠ *d* 15° 45′ . .	9·4505117

From 90° — 15° 45′ = 74° 15′, ∠ *ad*D.

And $\sqrt{\overline{6{\cdot}45}^2 + \overline{1{\cdot}82}^2} = 6{\cdot}7$ *d*D, or 6·70 chains.

Therefore the magnitude of the error, or the bearing and distance of D from *d*, will (from Art. 3) be N 74° 15′ E 6·70 chains with the true meridian.

Example III.—If the following subterraneous survey S 30° W 4 chains, N 50° W 8 chains, N 50° E 9 chains, and N 53° W 8 chains, is surveyed by an instrument having 23° of west variation, and plotted on a plan by the true meridian, without being reduced thereto,—what will be magnitude of the error thereby?

Suppose A to be the point of commencement on the plan, and NS the true meridian thereon; then ABCDF will be the erroneous representation of the bearings and distances, as plotted from that meridian,—AB forming an angle of 30° therewith, BC an angle of 50° therewith, CD an angle of 50° therewith, and DF an angle of 53° therewith.

To plot the survey accurately, draw on the plan a meridian line *ns*, having 23° of west variation; each bearing and distance being then plotted from it, and *Abcdf* will represent the survey accurately done, and *f* F will be the magnitude of the error: Or, otherwise, if each bearing in the survey is reduced from the angle it formed with the magnetic meridian it was taken by, to the angle of bearing it will form with the plan's meridian, which is the true meridian, and plotted accordingly, the result will be the same: Thus,—

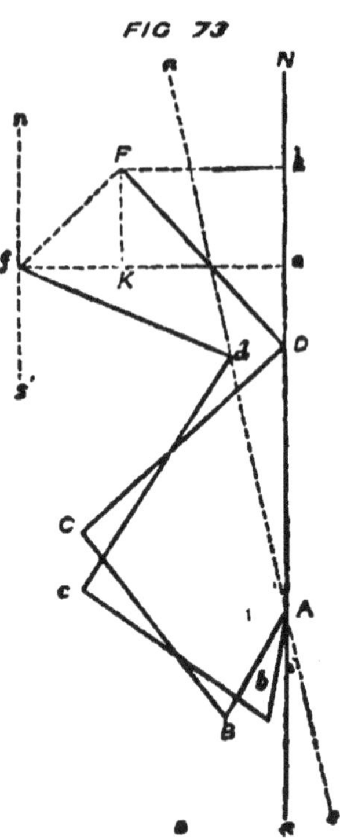

With a meridian having 23° of west variation.	Chains.	*With the true meridian.*	Chains.
S. 30° W. . .	4	S. 7° W. . .	4
N. 50° W. .	8	N. 73° W. .	8
N. 50° E. . .	9	N. 27° E. . .	9
N. 53° W. .	8	N. 76° W. .	8

Then A*b* will represent S 7° W 4 chains, *bc* N 73° W 8 chains, *cd* N 27° E 9 chains, and *df*, N 76° W 8 chains, the same as before.

	Northing.	Southing.	Easting.	Westing.	
Chns.	Chains.	Chains.	Chains.	Chains.	
S. 30° W. 4	...	3·46	...	2·00	
N. 50° W. 8	5·14	...	...	6·13	
N. 50° E. 9	5·79	...	6·89	...	
N. 53° W. 8	4·81	...	...	6·39	
	15·74			14·52	
	3·46			6·89	
	12·28	A*h*		7·63	*h*F or *ak*

	Northing.	Southing.	Easting.	Westing.	
Chns.	Chains.	Chains.	Chains.	Chains.	
S. 7° W. 4	...	3·97	...	0·48	
N. 73° W. 8	2·33	...	...	7·65	
N. 27° E. 9	8·01	...	4·08	...	
N. 76° W. 8	1·93		...	7·76	
	12·27			15·89	
	3·97			4·08	
	8·30	A*a*		11·81	*af*

From *af* 11·81 — *ak* 7·63 = *kf* 4·18.
A*h* 12·28 — A*a* 8·30 = *ah* or *k*F 3·98.

Then, as *kf* 4·18	·6211763
Is to radius	10·0000000
So is *k*F 3·98 . . .	5998831
To tang. ∠ *f* 43° 35′ . .	9·9787068

From 90° — 43° 35′ = 46° 25′ ∠ *nf*F.

And $\sqrt{\overline{4{\cdot}18}^2 + \overline{3{\cdot}98}^2} = 5{\cdot}77$ *f*F chains.

Therefore the bearing of F from *f* with the true meridian will be N 46° 25′ E, and the distance will be 5·77 chains; which is the magnitude of the error.

Example IV.—If the following subterraneous survey,

commencing at the pit A, S 30° W 4 chains, S 70° W 10 chains, and S 50° E 5 chains, was surveyed by an instrument whose needle had 23° of west variation, and is plotted on a plan by a meridian having only 10° of variation to the west, without reducing the bearings thereto; what will be magnitude of the error?

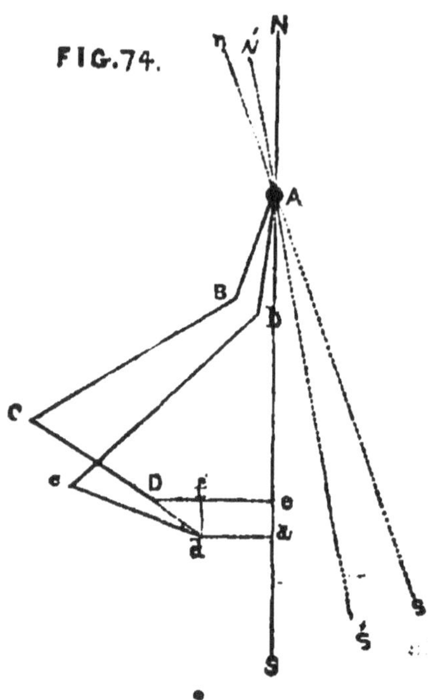

If NS represent the true meridian,—*ns* the meridian, having 23° of west variation, by which the survey was taken,—and *N'S'* the meridian of the plan, having 10° of variation, by which the survey is to be plotted; the ABCD will be the erroneous representation of the survey, as plotted by the meridian *N'S'* without reducing the bearings thereto. To plot the same truly,

With a meridian of 23° of west variation.	Chains.	*With a meridian of 10° of west variation.*	Chains.
S. 30° W.	4	S. 17° W.	4
S. 70° W.	10	S. 57° W.	10
S. 50° E.	5	S. 63° E.	5

Now make *Ab* form an angle to the west with the meridian *N'S'* of 17°, *bc* an angle to the west of 57°, and *cd* an angle to the east of 63°; then *Abcd* will represent the survey truly plotted, and the distance between D and *d* will be the magnitude of the error.

TO FIND THE MAGNITUDE OF THE ERROR.

With a meridian of 23° of west variation.	With the true meridian.	With a meridian of 10° of west variation.	With the true meridian.
Chns.	Chns.	Chns.	Chns.
S. 30° W. 4	S. 7° W. 4	S. 30° W. 4	S. 20° W. 4
S. 70° W. 10	S. 47° W. 10	S. 70° W. 10	S. 60° W. 10
S. 50° E. 5	S. 73° E. 5	S. 50° E. 5	S. 60° E. 5

	Northing.	Southing.	Easting.	Westing.	
Chns.	Chains.	Chains.	Chains.	Chains.	
S. 7° W. 4	...	3·79	...	0·49	
S. 47° W. 10	...	6·82	...	7·31	
S. 73° E. 5	...	1·46	4·78	...	
		12·25	*Aa*	7·80	
				4·78	
				3·02	*ad*

	Northing.	Southing.	Easting.	Westing.	
Chns.	Chains.	Chains.	Chains.	Chains.	
S. 20° W. 4	...	3·75	...	1·37	
S. 60° W. 10	...	5·00	...	8·66	
S. 60° E. 5	...	2·50	4·33	...	
		11·25	*Ae*	10·03	
				4·33	
				5·70	*e*D

Then A*a* 12·25 — A*e* 11·25 = *ae* or *fd* 1.
And *e*D 5·70 — *ad* 3·02 = *f*D 2·68.

As *f*D 2·68	·4281348
Is to radius	10·0000000
So is *fd* 1	
To tang. ∠ D. 20° 27′ . .	9·5718652

From 90° — 20° 27′ = 69° 33′, ∠ *fd*D.

And $\sqrt{2{\cdot}68^2 + 1^2}$ = 2·86 = D*d*.

Therefore the bearing of D from *d* with the true meridian will be N 69° 33′ W, and the distance will be 2·86 chains; which is the magnitude of the error.

EXAMPLE V.—The following subterraneous survey,—S 20° W 5 chains, S 70° W 10 chains, N 50° W 5 chains, and N 3° W 8 chains, was taken by an instrument having 23° of west variation, which I have to plot on a plan, the magnetic variation of the meridian by which it has been constructed is unknown; I therefore wish to know how the survey must be plotted, so that it may be accurately done?

In order to find by what kind of meridian the plan has been constructed, I took the bearing of two pits thereon by the delineated meridian, which I found to bear with each other N 25° W,—and the same two pits on the surface I found to bear N 22° W by an instrument whose needle had 23° of west variation; therefore the plan's meridian will have 20° of west variation, and the bearings of the survey must be reduced from a meridian of 23° of west variation to bearings with a meridian of 20° of the same variation, and plotted on the plan accordingly; Thus,—

Bearings with a meridian of 23° of west variation.	Chains.	*Bearings with the plan's meridian of 20° of west variation.*	Chains.
S. 20° W.	5	S. 17° W.	5
S. 70° W.	10	S. 67° W.	10
N. 50° W.	5	N. 53° W.	5
N. 3° W.	8	N. 6° W.	8

How to run bearings on the surface by a circumferentor, without error.

(56.) It frequently happens that the practical miner has to re-traverse on the surface the survey of a subterraneous excavation from bearings taken at some former time: Now, when that is the case, if the miner does it without attending to the change that has taken place with the magnetic meridian, between the taking of the survey and the retraversing it, an error must inevitably be the result; but where surveys are recorded without mentioning by what kind of meridian they were originally made, such surveys cannot be re-traversed with any degree of accuracy.

Suppose the bearing of a subterraneous excavation AB, is found to be N 20° W, which is taken by the needle of an instrument placed at the pit A, whose magnetic meridian is represented by NS; now, if the bearing of this excavation is run off on the surface from the pit A, immediately after it has been surveyed under-ground, and by the same instrument also, the excavation AB will be truly represented on the surface (see theorem 6, Art. 48); but if it should be necessary, at any future time, to have the same excavation represented on the surface by the same survey already made, and in that interval of time between the survey being made and its second plotting on the surface, the magnetic meridian NS has changed its situation to *ns*, the same excavation, N 20° W, run off from the then magnetic meridian *ns*, will be represented by A*b*, which will be erroneous: Therefore, to do the work truly, the bearing of AB, as originally taken by the meridian NS, must be reduced to its bearing with the meridian *ns*, and plotted on the surface from it accordingly (see theorem 7, Art. 48).

I shall insert a few examples relative to plotting bearings on the surface by different meridians.

EXAMPLE I.—The subterraneous excavation commencing at the pit A, N 20° W 5 chains AB, N 20° E 8 chains BC, N 70° E 5 chains CD, and S 70° E 5 chains DF, was surveyed by an instrument whose needle had 10° of west variation *ns*, and is to be plotted on the surface by another instrument whose needle has a different magnetic variation; how must it be plotted with accuracy?

First, find the magnetic variation of the needle of the instrument by which the survey is to be plotted (see Art. 51), which suppose it to have 23° of west variation N S; then reduce the bearings, as taken by a meridian of 10° of west variation *ns*, to bearings with a meridian of 23° of west variation N S.

Example II.—If the following survey of a subterraneous excavation, commencing at the pit A (see Fig. to Ex. IV. Art. 55), S 30° W 4 chains, S 70° W 10 chains, and S 50° E 5 chains, was surveyed by an instrument which had 10° of west variation; what will be the magnitude of the error, if the survey is plotted on the surface by another instrument having 23° of west variation?

Let N′S′ represent the magnetic meridian of the needle of the instrument by which the survey was made, having 10° of west variation, and let ABCD represent the survey as plotted on the surface thereby,—also let *ns* represent the meridian of the instrument whose needle has 23° of west variation, and *Abcd* the excavation as plotted according to that meridian; then ABCD will be the survey plotted truly, and *Abcd* the same plotted erroneously: Therefore, from the manner of determining the magnitude of an error, arising from plotting a survey by a different meridian than that by which it was made (Art. 55), the error will be 2·86 chains,—which is the distance of *d* from D.

Example III.—I have the survey of a subterraneous excavation, commencing at a pit called the A pit; the bearings are recorded to be taken by the true meridian, viz., N 20° W 5 chains, due north 8 chains, N 80° E 5 chains, N 45½° W 10 chains, and N 23½° W 4 chains; how is the survey to be truly delineated by an instrument on the surface, so that a pit may be sunk on the extreme point of the last bearing?

The first thing to be done, the surveyor must ascertain the magnetic variation of the needle of the instrument by which he intends delineating the survey (see Art. 49) which suppose to be 23° 30′ to the west, and reduce the bearings of the survey thereto: Thus,—

Bearings with the true meridian.	Chains.	*Bearings with a meridian of 25° 30′ of west variation.*	Chains.
N. 30° W. . . .	5	N. 6½° W. . . .	5
N.	8	N. 23½° E. . . .	8
N. 80° E. . . .	5	N. 76½° E. . . .	5
N. 45½° W. . . .	10	N. 22° W. . . .	10
N. 23½° W. . . .	4	N.	4

Then fix the instrument at the A pit, and run off the first bearing and distance N 6½° W 5 chains, and the other following ones in regular order, and the end of the last N 4 chains, will be the place on the surface where the pit must be sunk, to hit the extreme point of the excavation.

To find the antiquity of a plan by its delineated meridian.

(57.) As the magnetic meridian has, for a great number of years past, been veering about to the west, hence plans constructed at different times must have their magnetic meridians of different variation; those that are of the most ancient construction will have their meridians more easterly than those of a more modern date. Should a plan be found to have been constructed by a meridian having 11° 15′ of east variation, it will be reasonable to suppose it has been made about the year 1576; for at that time the magnetic meridian had 11° 15′ of east variation (see Table, Art. 49): Or, if its meridian is found to have 20° of west variation, from the same principle it may be supposed to have been made about the year 1765.

Example I.—If a plan is found to have a magnetic meridian of 18° of west variation, in what year has it been constructed?

By looking in the table, Art. 49, it will appear to have been made about the year 1750.

EXAMPLE II.—I have a plan on which is a delineated meridian; I therefore wish to know in what year it has been made?

First find the magnetic variation of the meridian on the plan according to the rules for finding the same, Art. 54, which suppose to be 6° of east variation; then, by the table, Art. 49, it will appear to have been made about the year 1622.

The manner of recording subterraneous surveys.

(58.) As the necessity of recording surveys of subterraneous workings frequently occurs, I shall therefore show how the same ought to be recorded, so that they may answer the intended design: Thus,—

A recorded survey of a subterraneous excavation, taken June 10th, 1800, beginning at the centre of the A pit, in Blackburn colliery.

Each bearing being reduced to the true meridian.

	Chains.
N. 10° W.	5·50
N. 20° E.	4·20
N. 75° E.	10·10
E.	4·40
S. 71° E.	6·30
N. 50° E.	5·90

A recorded survey of a subterraneous working, taken November 21st, 1801, beginning at the centre of the Venture Pit, in Tanfield colliery.

Each bearing was taken by a needle having 23° of west variation, and recorded accordingly.

	Chains.
S. 50° W.	5·24
S. 30° W.	2·20
S. 86° W.	5·70
N. 40° W.	12·60

Now, either of these recorded surveys may be truly re-traversed on the surface of the earth, at any future time with accuracy, by an instrument whose magnetic needle may have any known variation whatever, by referring to Art. 56.

The nature and use of the Traverse Tables.

(59.) Thus, if it is required to know the northing and easting of N 18° E 56 links,—look in the tables under the degree answering to the bearing, and to the right, opposite 56 in the column of bearing lengths, will be found 53 links and 26 hundred parts of a link of northing, and 17 links and 30 hundred parts of a link of easting. As the bearing length is links, the northing and easting must be links and parts of a link; for in whatever denomination the bearing length is, in the same denomination must the integral part of the northing or southing and easting or westing be.

Also, if it is required to know the northing and easting of N 18° E 5·65 chains,—look in the table under the degree answering to the bearing, and opposite 5 chains in the bearing lengths will be found 4·76 chains of northing and 1·55 chains of easting; then, for the remaining 65 links, look opposite 65 in the same column of bearing lengths, and there will be found 61·82 links of northing, and 20·09 links of easting,—which, added to the former northing and easting, will make 5·3782, or nearly 5·38 chains of northing, for the whole northing,—and 1·7509 chains, or 1·75 chains nearly, for the whole easting.

Suppose, again, the southing and westing of S 86° W 98·20 chains is required,—look in the tables under the degree of the bearing, and the southing and westing will be thus:—

	Chains.		Chains.		Chains.	
For	98·00	there is	6·84	of southing and	97·76	of westing.
For	00·20		0· 1·40	of ditto	0·19·95	of ditto.
	98·20		6·85·40	of southing and	97·95·95	of westing.
		or,	6·85⅖	of southing and	97·96	of westing nearly.

If the southing and easting of S 18½° E 20 chains is required,—take the southing and easting of the bearing length under 18°, and also under 19°, in manner before shewn, and half their sum will be the southing and easting required; thus:—

	Chains.		Chains.		Chains.	
S. 18° E.	20	will have	19·02	of southing	6·18	of easting.
S. 19° E.	20	will have	18·91	of ditto.	6·51	of ditto.
			2)37·93		12·69	
			18·96	of southing	6·34	of easting.

Therefore S 18½° E 20 chains will have 18·96 chains of southing and 6·34 chains of easting.

Again, if the northing and westing of N 75½° W 10·35 chains is required,—

	Chains.		Chains.		Chains.	
N. 75° W.	10·35	will have	2·68·06	of northing	9·99·81	westing.
N. 76° W.	10·35	will have	2·50·47	of ditto.	10·08·96	ditto.
			2)5·18·53		20·08·77	
N. 75½° W.	10·35	will have	2·59·26	of northing	10·04·38	westing,
		or nearly	2·59¼	of northing	10·04⅜	westing.

If the northing and easting of N 14° 37′ E 18 chains be required, take the northing and easting of the bearing length under 14° and the same under 15°; take the difference of each, multiply the respective differences by the number of minutes, *i. e.* 37′, and divide the products by 60 (the number of minutes in a degree), subtract the first quotient from the northing, and add the second to the easting; and the sum and difference will be the northing and easting required; thus—

N. 14° E. 18 chains will have 17·47 of northing, and 4·35 of easting.
N. 15° E. „ „ 17·39 of ditto, „ 4·66 of ditto.

	·08 diff.	·31 diff.
	37	37
	60) 29·6	60)114·7
	5 nearly.	19 nearly.
	14·47	4·35
N. 14° 37′ E. 18 ch. will have	14·42 of northing, and	4·54 of easting.

The use of the Traverse Tables in reducing hypothenusal or inclined distances to horizontal distances.—(See Art. 45.)

(60.) When the table is used for the before-mentioned purpose, the column called bearing lengths represents the hypothenusal distance or longest side of a right-angled triangle, as CB; the column called N or S distance represents the horizontal distance AB; and the column called E or W distance represents the perpendicular AC.

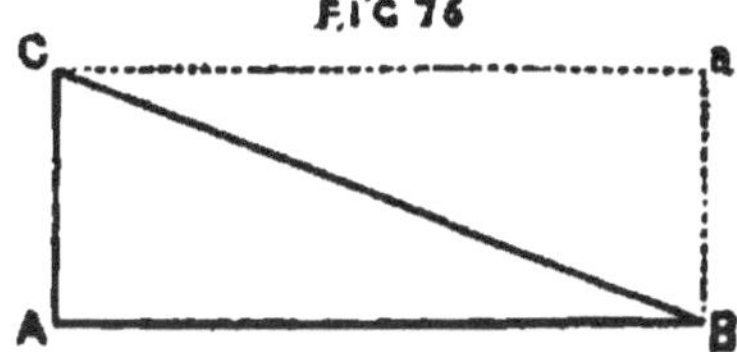

FIG 76

If the horizontal distance AB or C*a* is required, when the hypothenusal distance CB is 10 chains, and the angle *a*CB or CBA is 20°,—look in the table under 20°, and opposite 10, in the column of bearing lengths, will be found in the column of N or S distance 9·40, which will be 9·40 chains, equal to the horizontal distance AB or C*a*.

If the horizontal distance AB or C*a* is required, when the hypothenusal distance CB is 8 chains, and the angle *a*CB or CBA is 50°,—look in the tables under 50°, and opposite 8, in the column of bearing lengths, will be found 5·14 chains, in the column of N or S distance, which is equal to AB or C*a*, the horizontal distance.

The horizontal distance of a line 20·50 chains, run under an angle of 15° of elevation, is required?

Look in the tables under 15°, and in the column of

bearing lengths for 20·50 chains, the horizontal distance will be thus:—

	Chains.		Chains.	
For	20·00	of hyp. distance	19·32	of horizontal distance.
For	0·50	of hyp. distance	0·48	of horizontal distance.
For	20·50	of hyp. distance	19·80	the whole horizontal distance.

Therefore, 20·50 chains of hypothenusal or inclining length will be equal to 19 chains 80 links, or 19·80 chains of horizontal distance.

TRAVERSE TABLES;

OR,

TABLES OF THE NORTHING OR SOUTHING,

AND

EASTING OR WESTING;

WHEREIN THE DISTANCE IS EXTENDED TO ONE HUNDRED,
FOR EVERY DEGREE OF THE QUADRANT.

½°						1°					
Bearing Lengths.	N. or S. Distance.	E. or W. Distance.	Bearing Lengths.	N. or S. Distance.	E. or W. Distance.	Bearing Lengths.	N. or S. Distance.	E. or W. Distance.	Bearing Lengths.	N. or S. Distance.	E. or W. Distance.
1	1·00	0·01	51	51·00	0·45	1	1·00	0·02	51	50·99	0·89
2	2·00	0·02	52	52·00	0·45	2	2·00	0·03	52	51·99	0·91
3	3·00	0·03	53	53·00	0·46	3	3·00	0·05	53	52·99	0·92
4	4·00	0·03	54	54·00	0·47	4	4·00	0·07	54	53·99	0·94
5	5·00	0·04	55	55·00	0·48	5	5·00	0·09	55	54·99	0·96
6	6·00	0·05	56	56·00	0·49	6	6·00	0·10	56	55·99	0·98
7	7·00	0·06	57	57·00	0·50	7	7·00	0·12	57	56·99	0·99
8	8·00	0·07	58	58·00	0·51	8	8·00	0·14	58	57·99	1·01
9	9·00	0·08	59	59·00	0·52	9	9·00	0·16	59	58·99	1·03
10	10·00	0·09	60	60·00	0·52	10	10·00	0·17	60	59·99	1·05
11	11·00	0·10	61	61·00	0·53	11	11·00	0·19	61	60·99	1·07
12	12·00	0·10	62	62·00	0·54	12	12·00	0·21	62	61·99	1·09
13	13·00	0·11	63	63·00	0·54	13	13·00	0·23	63	62·99	1·10
14	14·00	0·12	64	64·00	0·55	14	14·00	0·24	64	63·99	1·12
15	15·00	0·13	65	65·00	0·56	15	15·00	0·26	65	64·99	1·14
16	16·00	0·13	66	66·00	0·57	16	16·00	0·28	66	65·99	1·16
17	17·00	0·14	67	67·00	0·58	17	17·00	0·29	67	66·99	1·17
18	18·00	0·15	68	68·00	0·59	18	18·00	0·31	68	67·99	1·19
19	19·00	0·16	69	69·00	0·60	19	19·00	0·33	69	68·99	1·21
20	20·00	0·17	70	70·00	0·61	20	20·00	0·35	70	69·99	1·22
21	21·00	0·18	71	71·00	0·62	21	21·00	0·37	71	70·99	1·24
22	22·00	0·18	72	72·00	0·63	22	22·00	0·38	72	71·99	1·26
23	23·00	0·19	73	73·00	0·63	23	23·00	0·40	73	72·99	1·28
24	24·00	0·20	74	74·00	0·64	24	24·00	0·42	74	73·99	1·29
25	25·00	0·21	75	75·00	0·65	25	25·00	0·44	75	74·99	1·31
26	26·00	0·22	76	76·00	0·66	26	26·00	0·45	76	75·99	1·33
27	27·00	0·23	77	77·00	0·67	27	27·00	0·47	77	76·99	1·35
28	28·00	0·24	78	78·00	0·68	28	28·00	0·49	78	77·99	1·36
29	29·00	0·25	79	79·00	0·69	29	29·00	0·51	79	78·99	1·38
30	30·00	0·26	80	80·00	0·70	30	30·00	0·52	80	79·99	1·40
31	31·00	0·26	81	81·00	0·71	31	31·00	0·54	81	80·99	1·42
32	32·00	0·27	82	82·00	0·72	32	32·00	0·56	82	81·99	1·44
33	33·00	0·28	83	83·00	0·73	33	33·00	0·58	83	82·99	1·45
34	34·00	0·29	84	84·00	0·74	34	33·99	0·60	84	83·99	1·47
35	35·00	0·30	85	85·00	0·74	35	34·99	0·61	85	84·99	1·49
36	36·00	0·31	86	86·00	0·75	36	35·99	0·63	86	85·99	1·51
37	37·00	0·32	87	87·00	0·76	37	36·99	0·65	87	86·99	1·53
38	38·00	0·33	88	88·00	0·77	38	37·99	0·67	88	87·99	1·54
39	39·00	0·34	89	89·00	0·78	39	38·99	0·69	89	88·99	1·56
40	40·00	0·35	90	90·00	0·79	40	39·99	0·70	90	89·99	1·57
41	41·00	0·36	91	91·00	0·80	41	40·99	0·72	91	90·99	1·59
42	42·00	0·36	92	92·00	0·81	42	41·99	0·74	92	91·99	1·61
43	43·00	0·37	93	93·00	0·81	43	42·99	0·76	93	92·99	1·62
44	44·00	0·38	94	94·00	0·82	44	43·99	0·78	94	93·99	1·64
45	45·00	0·39	95	95·00	0·83	45	44·99	0·79	95	94·99	1·66
46	46·00	0·40	96	96·00	0·84	46	45·99	0·81	96	95·99	1·68
47	47·00	0·41	97	97·00	0·85	47	46·99	0·83	97	96·99	1·69
48	48·00	0·42	98	98·00	0·85	48	47·99	0·84	98	97·99	1·71
49	49·00	0·43	99	99·00	0·86	49	48·99	0·86	99	98·99	1·73
50	50·00	0·44	100	100·00	0·87	50	49·99	0·87	100	99·98	1·75
	E. or W.	N. or S.		E. or W.	N. or S.		E. or W.	N. or S.		E. or W.	N. or S.
89½°						89°					

2°						3°					
Bearing Lengths.	N. or S. Distance.	E. or W. Distance.	Bearing Lengths.	N. or S. Distance.	E. or W. Distance.	Bearing Lengths.	N. or S. Distance.	E. or W. Distance.	Bearing Lengths.	N. or S. Distance.	E. or W. Distance.
1	1·00	0·03	51	50·97	1·78	1	1·00	0·05	51	50·93	2·67
2	2·00	0·07	52	51·97	1·81	2	2·00	0·11	52	51·93	2·72
3	3·00	0·10	53	52·97	1·85	3	3·00	0·16	53	52·93	2·77
4	4·00	0·14	54	53·97	1·88	4	3·99	0·21	54	53·93	2·83
5	5·00	0·17	55	54·97	1·92	5	4·99	0·26	55	54·93	2·88
6	6·00	0·21	56	55·97	1·95	6	5·99	0·31	56	55·92	2·93
7	7·00	0·24	57	56·97	1·99	7	6·99	0·37	57	56·92	2·98
8	8·00	0·28	58	57·97	2·02	8	7·99	0·42	58	57·92	3·04
9	8·99	0·31	59	58·96	2·06	9	8·99	0·47	59	58·92	3·09
10	9·99	0·35	60	59·96	2·09	10	9·99	0·52	60	59·92	3·14
11	10·99	0·38	61	60·96	2·13	11	10·98	0·58	61	60·92	3·19
12	11·99	0·42	62	61·96	2·16	12	11·98	0·63	62	61·92	3·25
13	12·99	0·45	63	62·96	2·20	13	12·98	0·68	63	62·92	3·30
14	13·99	0·49	64	63·96	2·23	14	13·98	0·73	64	63·91	3·35
15	14·99	0·52	65	64·96	2·27	15	14·98	0·79	65	64·91	3·40
16	15·99	0·56	66	65·96	2·30	16	15·98	0·84	66	65·91	3·46
17	16·99	0·59	67	66·96	2·34	17	16·98	0·89	67	66·91	3·51
18	17·99	0·63	68	67·96	2·37	18	17·98	0·94	68	67·91	3·56
19	18·99	0·66	69	68·96	2·40	19	18·97	1·00	69	68·91	3·61
20	19·99	0·70	70	69·96	2·44	20	19·97	1·05	70	69·90	3·66
21	20·99	0·73	71	70·96	2·47	21	20·97	1·10	71	70·90	3·72
22	21·99	0·77	72	71·96	2·51	22	21·97	1·15	72	71·90	3·77
23	22·98	0·80	73	72·96	2·54	23	22·97	1·20	73	72·90	3·82
24	23·98	0·84	74	73·95	2·58	24	23·97	1·26	74	73·90	3·88
25	24·98	0·87	75	74·95	2·61	25	24·97	1·31	75	74·90	3·93
26	25·98	0·91	76	75·95	2·65	26	25·96	1·36	76	75·90	3·98
27	26·98	0·94	77	76·95	2·68	27	26·96	1·41	77	76·90	4·04
28	27·98	0·98	78	77·95	2·72	28	27·96	1·47	78	77·89	4·09
29	28·98	1·01	79	78·95	2·75	29	28·96	1·52	79	78·89	4·14
30	29·98	1·05	80	79·95	2·79	30	29·96	1·57	80	79·89	4·19
31	30·98	1·08	81	80·95	2·82	31	30·96	1·62	81	80·89	4·24
32	31·98	1·12	82	81·95	2·86	32	31·96	1·68	82	81·89	4·29
33	32·98	1·15	83	82·95	2·89	33	32·95	1·73	83	82·89	4·35
34	33·98	1·19	84	83·95	2·93	34	33·95	1·78	84	83·89	4·40
35	34·98	1·22	85	84·95	2·96	35	34·95	1·83	85	84·88	4·45
36	35·98	1·26	86	85·95	3·00	36	35·95	1·88	86	85·88	4·50
37	36·98	1·29	87	86·95	3·03	37	36·95	1·94	87	86·88	4·56
38	37·98	1·33	88	87·95	3·07	38	37·95	1·99	88	87·88	4·61
39	38·98	1·36	89	88·95	3·10	39	38·95	2·04	89	88·88	4·66
40	39·98	1·40	90	89·95	3·14	40	39·95	2·09	90	89·88	4·71
41	40·98	1·43	91	90·94	3·17	41	40·94	2·15	91	90·88	4·76
42	41·98	1·47	92	91·94	3·21	42	41·94	2·20	92	91·87	4·82
43	42·98	1·50	93	92·94	3·24	43	42·94	2·25	93	92·87	4·87
44	43·97	1·53	94	93·94	3·28	44	43·94	2·30	94	93·87	4·92
45	44·97	1·57	95	94·94	3·31	45	44·94	2·36	95	94·87	4·97
46	45·97	1·60	96	95·94	3·35	46	45·94	2·41	96	95·87	5·02
47	46·97	1·64	97	96·94	3·38	47	46·94	2·46	97	96·87	5·08
48	47·97	1·67	98	97·94	3·42	48	47·94	2·51	98	97·87	5·13
49	48·97	1·71	99	98·94	3·45	49	48·93	2·57	99	98·87	5·18
50	49·97	1·74	100	99·94	3·49	50	49·93	2·62	100	99·86	5·23
	E. or W.	N. or S.		E. or W.	N. or S.		E. or W.	N. or S.		E. or W.	N. or S.
88°						87°					

4°					
Bearing Lengths	N. or S. Distance.	E. or W. Distance.	Bearing Lengths.	N. or S. Distance.	E. or W. Distance.
1	1·00	0·07	51	50·88	3·56
2	2·00	0·14	52	51·87	3·63
3	2·99	0·21	53	52·87	3·70
4	3·99	0·28	54	53·87	3·77
5	4·99	0·35	55	54·87	3·84
6	5·99	0·42	56	55·86	3·91
7	6·98	0·49	57	56·86	3·98
8	7·98	0·56	58	57·86	4·05
9	8·98	0·63	59	58·86	4·12
10	9·98	0·70	60	59·85	4·19
11	10·97	0·77	61	60·85	4·26
12	11·97	0·84	62	61·85	4·32
13	12·97	0·91	63	62·85	4·39
14	13·97	0·98	64	63·84	4·46
15	14·96	1·05	65	64·84	4·53
16	15·96	1·12	66	65·84	4·60
17	16·96	1·19	67	66·84	4·67
18	17·96	1·26	68	67·83	4·74
19	18·95	1·33	69	68·83	4·81
20	19·95	1·40	70	69·83	4·88
21	20·95	1·47	71	70·83	4·95
22	21·95	1·54	72	71·82	5·02
23	22·94	1·61	73	72·82	5·09
24	23·94	1·68	74	73·82	5·16
25	24·94	1·75	75	74·82	5·23
26	25·94	1·82	76	75·81	5·30
27	26·93	1·89	77	76·81	5·37
28	27·93	1·96	78	77·81	5·44
29	28·93	2·03	79	78·81	5·51
30	29·93	2·09	80	79·81	5·58
31	30·92	2·16	81	80·80	5·65
32	31·92	2·23	82	81·80	5·72
33	32·92	2·30	83	82·80	5·79
34	33·92	2·37	84	83·80	5·86
35	34·91	2·44	85	84·79	5·93
36	35·91	2·51	86	85·79	6·00
37	36·91	2·58	87	86·79	6·07
38	37·91	2·65	88	87·79	6·14
39	38·90	2·72	89	88·78	6·21
40	39·90	2·79	90	89·78	6·28
41	40·90	2·86	91	90·78	6·35
42	41·90	2·93	92	91·78	6·42
43	42·90	3·00	93	92·77	6·49
44	43·89	3·07	94	93·77	6·56
45	44·89	3·14	95	94·77	6·63
46	45·89	3·21	96	95·77	6·70
47	46·89	3·28	97	96·76	6·77
48	47·88	3·35	98	97·76	6·84
49	48·88	3·42	99	98·76	6·91
50	49·88	3·49	100	99·76	6·98
	E. or W.	N. or S.		E. or W.	N. or S.
86°					

5°					
Bearing Lengths.	N. or S. Distance.	E. or W. Distance.	Bearing Lengths.	N. or S. Distance.	E. or W. Distance.
1	1·00	0·09	51	50·81	4·45
2	1·99	0·17	52	51·80	4·53
3	2·99	0·26	53	52·80	4·62
4	3·98	0·35	54	53·79	4·71
5	4·98	0·44	55	54·79	4·79
6	5·98	0·52	56	55·79	4·88
7	6·97	0·61	57	56·78	4·97
8	7·97	0·70	58	57·78	5·06
9	8·97	0·78	59	58·78	5·14
10	9·96	0·87	60	59·77	5·23
11	10·96	0·96	61	60·77	5·32
12	11·95	1·05	62	61·76	5·41
13	12·95	1·13	63	62·76	5·49
14	13·95	1·22	64	63·76	5·58
15	14·94	1·31	65	64·75	5·67
16	15·94	1·39	66	65·75	5·75
17	16·94	1·48	67	66·75	5·84
18	17·93	1·57	68	67·74	5·93
19	18·93	1·66	69	68·74	6·02
20	19·92	1·74	70	69·73	6·10
21	20·92	1·83	71	70·73	6·19
22	21·92	1·92	72	71·73	6·28
23	22·91	2·00	73	72·72	6·36
24	23·91	2·09	74	73·72	6·45
25	24·91	2·18	75	74·72	6·54
26	25·90	2·27	76	75·71	6·63
27	26·90	2·35	77	76·71	6·71
28	27·89	2·44	78	77·70	6·80
29	28·89	2·53	79	78·70	6·89
30	29·89	2·61	80	79·70	6·97
31	30·88	2·70	81	80·69	7·06
32	31·88	2·79	82	81·69	7·15
33	32·88	2·88	83	82·68	7·24
34	33·87	2·96	84	83·68	7·32
35	34·87	3·05	85	84·68	7·41
36	35·86	3·14	86	85·67	7·50
37	36·86	3·22	87	86·67	7·58
38	37·86	3·31	88	87·67	7·67
39	38·85	3·40	89	88·66	7·76
40	39·85	3·49	90	89·66	7·84
41	40·84	3·57	91	90·65	7·93
42	41·84	3·66	92	91·65	8·02
43	42·84	3·75	93	92·65	8·11
44	43·83	3·84	94	93·64	8·19
45	44·83	3·92	95	94·64	8·28
46	45·83	4·01	96	95·64	8·37
47	46·82	4·10	97	96·63	8·45
48	47·82	4·18	98	97·63	8·54
49	48·81	4·27	99	98·62	8·63
50	49·81	4·36	100	99·62	8·72
	E. or W.	N. or S.		E. or W.	N. or S.
85°					

6°						7°					
Bearing Lengths.	N. or S. Distance.	E. or W. Distance.	Bearing Lengths.	N. or S. Distance.	E. or W. Distance.	Bearing Lengths.	N. or S. Distance.	E. or W. Distance.	Bearing Lengths.	N. or S. Distance.	E. or W. Distance.
1	0·99	0·10	51	50·72	5·33	1	0·99	0·12	51	50·62	6·22
2	1·99	0·21	52	51·72	5·44	2	1·99	0·24	52	51·61	6·34
3	2·98	0·31	53	52·71	5·54	3	2·98	0·37	53	52·60	6·46
4	3·98	0·42	54	53·70	5·64	4	3·97	0·49	54	53·60	6·58
5	4·97	0·52	55	54·70	5·75	5	4·96	0·61	55	54·59	6·70
6	5·97	0·63	56	55·69	5·85	6	5·96	0·73	56	55·58	6·82
7	6·96	0·73	57	56·69	5·96	7	6·94	0·85	57	56·57	6·95
8	7·96	0·84	58	57·68	6·06	8	7·94	0·97	58	57·57	7·07
9	8·95	0·94	59	58·68	6·17	9	8·93	1·10	59	58·56	7·19
10	9·95	1·05	60	59·67	6·27	10	9·93	1·22	60	59·55	7·31
11	10·94	1·15	61	60·67	6·38	11	10·92	1·34	61	60·54	7·43
12	11·93	1·25	62	61·66	6·48	12	11·91	1·46	62	61·54	7·56
13	12·93	1·36	63	62·65	6·59	13	12·90	1·58	63	62·53	7·68
14	13·92	1·46	64	63·65	6·69	14	13·90	1·71	64	63·52	7·80
15	14·92	1·57	65	64·64	6·79	15	14·89	1·83	65	64·51	7·92
16	15·91	1·67	66	65·64	6·90	16	15·88	1·95	66	65·51	8·04
17	16·91	1·78	67	66·63	7·00	17	16·87	2·07	67	66·50	8·17
18	17·90	1·88	68	67·63	7·11	18	17·87	2·19	68	67·49	8·29
19	18·90	1·99	69	68·62	7·21	19	18·86	2·32	69	68·48	8·41
20	19·89	2·09	70	69·62	7·32	20	19·85	2·44	70	69·48	8·53
21	20·88	2·20	71	70·61	7·42	21	20·84	2·56	71	70·47	8·65
22	21·88	2·30	72	71·61	7·53	22	21·84	2·68	72	71·46	8·77
23	22·87	2·41	73	72·60	7·63	23	22·83	2·80	73	72·45	8·90
24	23·87	2·51	74	73·59	7·74	24	23·82	2·92	74	73·45	9·02
25	24·86	2·61	75	74·59	7·84	25	24·81	3·05	75	74·44	9·14
26	25·86	2·72	76	75·58	7·94	26	25·81	3·17	76	75·43	9·26
27	26·85	2·82	77	76·58	8·05	27	26·80	3·29	77	76·42	9·38
28	27·85	2·93	78	77·57	8·15	28	27·79	3·41	78	77·42	9·51
29	28·84	3·03	79	78·57	8·26	29	28·78	3·53	79	78·41	9·63
30	29·84	3·14	80	79·56	8·36	30	29·78	3·66	80	79·40	9·75
31	30·83	3·24	81	80·55	8·47	31	30·77	3·78	81	80·39	9·87
32	31·82	3·34	82	81·55	8·57	32	31·76	3·90	82	81·39	9·99
33	32·82	3·45	83	82·55	8·68	33	32·75	4·02	83	82·38	10·12
34	33·81	3·56	84	83·54	8·78	34	33·75	4·14	84	83·37	10·24
35	34·81	3·66	85	84·53	8·89	35	34·74	4·27	85	84·36	10·36
36	35·80	3·76	86	85·53	8·99	36	35·73	4·39	86	85·36	10·48
37	36·80	3·87	87	86·52	9·10	37	36·72	4·51	87	86·35	10·60
38	37·79	3·97	88	87·52	9·20	38	37·72	4·63	88	87·34	10·72
39	38·79	4·08	89	88·51	9·31	39	38·71	4·75	89	88·33	10·85
40	39·78	4·18	90	89·51	9·41	40	39·70	4·87	90	89·33	10·97
41	40·77	4·29	91	90·50	9·52	41	40·69	5·00	91	90·32	11·09
42	41·77	4·39	92	91·50	9·62	42	41·69	5·12	92	91·31	11·21
43	42·76	4·49	93	92·49	9·72	43	42·68	5·24	93	92·31	11·33
44	43·76	4·60	94	93·48	9·83	44	43·67	5·36	94	93·30	11·46
45	44·75	4·70	95	94·48	9·93	45	44·66	5·48	95	94·29	11·58
46	45·75	4·81	96	95·47	10·04	46	45·66	5·61	96	95·28	11·70
47	46·74	4·91	97	96·47	10·14	47	46·65	5·73	97	96·28	11·82
48	47·74	5·02	98	97·46	10·25	48	47·64	5·85	98	97·27	11·94
49	48·73	5·12	99	98·46	10·35	49	48·63	5·97	99	98·26	12·07
50	49·73	5·23	100	99·45	10·45	50	49·63	6·09	100	99·26	12·19
	E. or W.	N. or S.		E. or W.	N. or S.		E. or W.	N. or S.		E. or W.	N. or S.
84°						83°					

8°						9°					
Bearing Lengths	N. or S. Distance.	E. or W. Distance.	Bearing Lengths.	N. or S. Distance.	E. or W. Distance.	Bearing Lengths.	N. or S. Distance.	E. or W. Distance.	Bearing Lengths.	N. or S. Distance.	E. or W. Distance.
1	0·99	0·14	51	50·50	7·10	1	0·99	0·16	51	50·37	7·98
2	1·98	0·27	52	51·49	7·24	2	1·98	0·31	52	51·36	8·13
3	2·97	0·42	53	52·48	7·38	3	2·96	0·47	53	52·35	8·29
4	3·96	0·56	54	53·47	7·52	4	3·95	0·63	54	53·34	8·45
5	4·95	0·70	55	54·46	7·65	5	4·94	0·78	55	54·32	8·60
6	5·94	0·84	56	55·46	7·79	6	5·93	0·94	56	55·31	8·76
7	6·93	0·97	57	56·45	7·93	7	6·91	1·10	57	56·30	8·92
8	7·92	1·11	58	57·44	8·07	8	7·90	1·25	58	57·29	9·07
9	8·91	1·25	59	58·43	8·21	9	8·89	1·41	59	58·27	9·23
10	9·90	1·39	60	59·42	8·35	10	9·88	1·56	60	59·26	9·39
11	10·89	1·53	61	60·41	8·49	11	10·86	1·72	61	60·25	9·54
12	11·88	1·67	62	61·40	8·63	12	11·85	1·88	62	61·24	9·70
13	12·87	1·81	63	62·39	8·77	13	12·84	2·03	63	62·22	9·86
14	13·86	1·95	64	63·38	8·91	14	13·83	2·19	64	63·21	10·01
15	14·85	2·09	65	64·37	9·05	15	14·82	2·35	65	64·20	10·17
16	15·84	2·23	66	65·36	9·19	16	15·80	2·50	66	65·19	10·32
17	16·83	2·37	67	66·35	9·32	17	16·79	2·66	67	66·18	10·48
18	17·82	2·51	68	67·34	9·46	18	17·78	2·82	68	67·16	10·64
19	18·82	2·64	69	68·33	9·60	19	18·77	2·97	69	68·15	10·79
20	19·81	2·78	70	69·32	9·74	20	19·75	3·13	70	69·14	10·95
21	20·80	2·92	71	70·31	9·88	21	20·74	3·29	71	70·13	11·11
22	21·79	3·06	72	71·30	10·02	22	21·73	3·44	72	71·11	11·26
23	22·78	3·20	73	72·29	10·16	23	22·72	3·60	73	72·10	11·42
24	23·77	3·34	74	73·28	10·30	24	23·70	3·75	74	73·09	11·58
25	24·76	3·48	75	74·27	10·44	25	24·69	3·91	75	74·08	11·73
26	25·75	3·62	76	75·26	10·58	26	25·68	4·07	76	75·06	11·89
27	26·74	3·76	77	76·25	10·72	27	26·67	4·22	77	76·05	12·05
28	27·73	3·90	78	77·24	10·86	28	27·66	4·38	78	77·04	12·20
29	28·72	4·04	79	78·23	10·99	29	28·64	4·54	79	78·03	12·36
30	29·71	4·18	80	79·22	11·13	30	29·63	4·69	80	79·02	12·52
31	30·70	4·31	81	80·21	11·27	31	30·62	4·85	81	80·00	12·67
32	31·69	4·45	82	81·20	11·41	32	31·61	5·01	82	80·99	12·83
33	32·68	4·59	83	82·19	11·55	33	32·59	5·16	83	81·98	12·98
34	33·67	4·73	84	83·18	11·69	34	33·58	5·32	84	82·97	13·14
35	34·66	4·87	85	84·17	11·83	35	34·57	5·48	85	83·95	13·30
36	35·65	5·01	86	85·16	11·97	36	35·56	5·63	86	84·94	13·45
37	36·64	5·15	87	86·15	12·11	37	36·54	5·79	87	85·93	13·61
38	37·63	5·29	88	87·14	12·25	38	37·53	5·94	88	86·92	13·77
39	38·62	5·43	89	88·13	12·39	39	38·52	6·10	89	87·90	13·92
40	39·61	5·57	90	89·12	12·53	40	39·51	6·26	90	88·89	14·08
41	40·60	5·71	91	90·11	12·66	41	40·50	6·41	91	89·88	14·24
42	41·59	5·85	92	91·10	12·80	42	41·48	6·57	92	90·87	14·39
43	42·58	5·98	93	92·09	12·94	43	42·47	6·73	93	91·86	14·55
44	43·57	6·12	94	93·09	13·08	44	43·46	6·88	94	92·84	14·70
45	44·56	6·27	95	94·08	13·22	45	44·45	7·04	95	93·83	14·86
46	45·55	6·40	96	95·07	13·36	46	45·43	7·20	96	94·82	15·02
47	46·54	6·54	97	96·06	13·50	47	46·42	7·35	97	95·81	15·17
48	47·53	6·68	98	97·05	13·64	48	47·41	7·51	98	96·79	15·33
49	48·52	6·82	99	98·04	13·78	49	48·40	7·67	99	97·78	15·49
50	49·51	6·96	100	99·03	13·92	50	49·38	7·82	100	98·77	15·64
	E. or W	N. or S.		E. or W.	N. or S.		E. or W.	N. or S.		E. or W.	N. or S.
82°						81°					

10°					
Bearing Lengths.	N. or S. Distance.	E. or W. Distance.	Bearing Lengths.	N. or S. Distance.	E. or W. Distance.
1	0·98	0·17	51	50·23	8·86
2	1·97	0·35	52	51·21	9·03
3	2·95	0·52	53	52·19	9·20
4	3·94	0·70	54	53·18	9·38
5	4·92	0·87	55	54·16	9·55
6	5·91	1·04	56	55·15	9·72
7	6·89	1·22	57	56·13	9·90
8	7·88	1·39	58	57·12	10·07
9	8·86	1·56	59	58·10	10·25
10	9·85	1·74	60	59·09	10·42
11	10·83	1·91	61	60·07	10·59
12	11·82	2·08	62	61·06	10·77
13	12·80	2·26	63	62·04	10·94
14	13·79	2·43	64	63·03	11·11
15	14·77	2·60	65	64·01	11·29
16	15·76	2·78	66	65·00	11·46
17	16·74	2·95	67	65·98	11·63
18	17·73	3·12	68	66·97	11·81
19	18·71	3·30	69	67·95	11·98
20	19·70	3·47	70	68·94	12·16
21	20·68	3·65	71	69·92	12·33
22	21·67	3·82	72	70·91	12·50
23	22·65	3·99	73	71·89	12·68
24	23·64	4·17	74	72·88	12·85
25	24·62	4·34	75	73·86	13·02
26	25·60	4·51	76	74·85	13·20
27	26·59	4·69	77	75·83	13·37
28	27·57	4·86	78	76·82	13·54
29	28·56	5·04	79	77·80	13·72
30	29·54	5·21	80	78·78	13·89
31	30·53	5·38	81	79·77	14·07
32	31·51	5·56	82	80·75	14·24
33	32·50	5·73	83	81·74	14·41
34	33·48	5·90	84	82·72	14·59
35	34·47	6·08	85	83·71	14·76
36	35·45	6·25	86	84·69	14·93
37	36·44	6·43	87	85·68	15·11
38	37·42	6·60	88	86·66	15·28
39	38·41	6·77	89	87·65	15·45
40	39·39	6·95	90	88·63	15·63
41	40·38	7·12	91	89·62	15·80
42	41·36	7·29	92	90·60	15·98
43	42·35	7·47	93	91·59	16·15
44	43·33	7·64	94	92·57	16·32
45	44·32	7·81	95	93·56	16·50
46	45·30	7·99	96	94·54	16·67
47	46·29	8·16	97	95·53	16·84
48	47·27	8·34	98	96·51	17·02
49	48·26	8·51	99	97·50	17·19
50	49·24	8·68	100	98·48	17·37
	E. or W.	N. or S.		E. or W.	N. or S.
80°					

11°					
Bearing Lengths.	N. or S. Distance.	E. or W. Distance.	Bearing Lengths.	N. or S. Distance.	E. or W. Distance.
1	0·98	0·19	51	50·06	9·73
2	1·96	0·38	52	51·04	9·92
3	2·94	0·57	53	52·03	10·11
4	3·93	0·76	54	53·01	10·30
5	4·91	0·95	55	53·99	10·49
6	5·89	1·14	56	54·97	10·69
7	6·87	1·34	57	55·95	10·88
8	7·85	1·53	58	56·93	11·07
9	8·83	1·72	59	57·92	11·26
10	9·82	1·91	60	58·90	11·45
11	10·80	2·10	61	59·88	11·64
12	11·78	2·29	62	60·86	11·83
13	12·76	2·48	63	61·84	12·02
14	13·74	2·67	64	62·82	12·21
15	14·72	2·86	65	63·80	12·40
16	15·71	3·05	66	64·79	12·59
17	16·69	3·24	67	65·77	12·78
18	17·67	3·43	68	66·75	12·98
19	18·65	3·63	69	67·73	13·17
20	19·63	3·82	70	68·71	13·36
21	20·61	4·01	71	69·69	13·55
22	21·60	4·20	72	70·68	13·74
23	22·58	4·39	73	71·66	13·93
24	23·56	4·58	74	72·64	14·12
25	24·54	4·77	75	73·62	14·31
26	25·52	4·96	76	74·60	14·50
27	26·50	5·15	77	75·58	14·69
28	27·49	5·34	78	76·57	14·88
29	28·47	5·53	79	77·55	15·07
30	29·45	5·72	80	78·53	15·26
31	30·43	5·92	81	79·51	15·46
32	31·41	6·11	82	80·49	15·65
33	32·39	6·30	83	81·47	15·84
34	33·37	6·49	84	82·46	16·03
35	34·36	6·68	85	83·44	16·22
36	35·34	6·87	86	84·42	16·41
37	36·32	7·06	87	85·40	16·60
38	37·30	7·25	88	86·38	16·79
39	38·28	7·44	89	87·36	16·98
40	39·27	7·63	90	88·35	17·17
41	40·25	7·82	91	89·33	17·36
42	41·23	8·01	92	90·31	17·55
43	42·21	8·20	93	91·29	17·75
44	43·19	8·40	94	92·27	17·94
45	44·17	8·59	95	93·25	18·13
46	45·15	8·78	96	94·24	18·32
47	46·14	8·97	97	95·22	18·51
48	47·12	9·16	98	96·20	18·70
49	48·10	9·35	99	97·18	18·89
50	49·08	9·54	100	98·16	19·08
	E. or W.	N. or S.		E. or W.	N. or S.
79°					

12°						13°					
Bearing Lengths.	N. or S. Distance.	E. or W. Distance.	Bearing Lengths.	N. or S. Distance.	E. or W. Distance.	Bearing Lengths.	N. or S. Distance.	E. or W. Distance.	Bearing Lengths.	N. or S. Distance.	E. or W. Distance.
1	0·98	0·21	51	49·89	10·60	1	0·97	0·22	51	49·69	11·47
2	1·96	0·42	52	50·86	10·81	2	1·95	0·45	52	50·67	11·70
3	2·93	0·62	53	51·84	11·02	3	2·92	0·67	53	51·64	11·92
4	3·91	0·83	54	52·82	11·23	4	3·90	0·90	54	52·62	12·15
5	4·89	1·04	55	53·80	11·44	5	4·87	1·12	55	53·59	12·37
6	5·87	1·25	56	54·78	11·64	6	5·85	1·35	56	54·56	12·60
7	6·85	1·46	57	55·75	11·85	7	6·82	1·57	57	55·54	12·82
8	7·83	1·66	58	56·73	12·06	8	7·79	1·80	58	56·51	13·05
9	8·80	1·87	59	57·71	12·27	9	8·77	2·02	59	57·49	13·27
10	9·78	2·08	60	58·69	12·47	10	9·74	2·25	60	58·46	13·50
11	10·76	2·29	61	59·67	12·68	11	10·72	2·47	61	59·44	13·72
12	11·74	2·49	62	60·65	12·89	12	11·69	2·70	62	60·41	13·95
13	12·72	2·70	63	61·62	13·10	13	12·67	2·92	63	61·39	14·17
14	13·69	2·91	64	62·60	13·31	14	13·64	3·15	64	62·36	14·40
15	14·67	3·12	65	63·58	13·51	15	14·62	3·37	65	63·33	14·62
16	15·65	3·33	66	64·56	13·72	16	[illegible]	3·60	66	64·31	14·85
17	16·63	3·53	67	65·54	13·93	17	16·56	3·82	67	65·28	15·07
18	17·61	3·74	68	66·51	14·14	18	17·54	4·05	68	66·26	15·30
19	18·58	3·95	69	67·49	14·35	19	18·51	4·27	69	67·23	15·52
20	19·56	4·16	70	68·47	14·55	20	19·49	4·50	70	68·21	15·75
21	20·54	4·37	71	69·45	14·76	21	20·46	4·72	71	69·18	15·97
22	21·52	4·57	72	70·43	14·97	22	21·44	4·95	72	70·16	16·20
23	22·50	4·78	73	71·40	15·18	23	22·41	5·17	73	71·13	16·42
24	23·48	4·99	74	72·38	15·39	24	23·38	5·40	74	72·10	16·65
25	24·45	5·20	75	73·36	15·59	25	24·36	5·62	75	73·08	16·87
26	25·43	5·41	76	74·34	15·80	26	25·33	5·85	76	74·05	17·10
27	26·41	5·61	77	75·32	16·01	27	26·31	6·07	77	75·03	17·32
28	27·39	5·82	78	76·30	16·22	28	27·28	6·30	78	76·00	17·55
29	28·37	6·03	79	77·27	16·43	29	28·26	6·52	79	76·98	17·77
30	29·34	6·24	80	78·25	16·63	30	29·23	6·75	80	77·95	18·00
31	30·32	6·45	81	79·23	16·84	31	30·21	6·97	81	78·92	18·22
32	31·30	6·65	82	80·21	17·05	32	31·18	7·20	82	79·90	18·45
33	32·28	6·86	83	81·19	17·26	33	32·15	7·42	83	80·87	18·67
34	33·26	7·07	84	82·16	17·46	34	33·13	7·65	84	81·85	18·90
35	34·24	7·28	85	83·14	17·67	35	34·10	7·87	85	82·82	19·12
36	35·21	7·48	86	84·12	17·88	36	35·08	8·10	86	83·80	19·35
37	36·19	7·69	87	85·10	18·09	37	36·05	8·32	87	84·77	19·57
38	37·17	7·90	88	86·08	18·30	38	37·03	8·55	88	85·74	19·80
39	38·15	8·11	89	87·06	18·50	39	38·00	8·77	89	86·72	20·02
40	39·13	8·32	90	88·03	18·71	40	38·97	9·00	90	87·69	20·25
41	40·10	8·52	91	89·01	18·92	41	39·95	9·22	91	88·67	20·47
42	41·08	8·73	92	89·99	19·13	42	40·92	9·45	92	89·64	20·70
43	42·06	8·94	93	90·97	19·34	43	41·90	9·67	93	90·62	20·92
44	43·04	9·15	94	91·95	19·54	44	42·87	9·90	94	91·59	21·15
45	44·02	9·36	95	92·92	19·75	45	43·85	10·12	95	92·57	21·37
46	44·99	9·56	96	93·90	19·96	46	44·82	10·35	96	93·54	21·60
47	45·97	9·77	97	94·88	20·17	47	45·80	10·57	97	94·51	21·82
48	46·95	9·98	98	95·86	20·38	48	46·77	10·80	98	95·49	22·05
49	47·93	10·19	99	96·84	20·58	49	47·74	11·02	99	96·46	22·27
50	48·91	10·40	100	97·81	20·79	50	48·72	11·25	100	97·44	22·50
	E. or W.	N. or S.		E. or W.	N. or S.		E. or W.	N. or S.		E. or W.	N. or S.
78°						77°					

14°					15°					
N. or S. Distance.	E. or W. Distance.	Bearing Lengths.	N. or S. Distance.	E. or W. Distance.	Bearing Lengths.	N. or S. Distance.	E. or W. Distance.	Bearing Lengths.	N. or S. Distance.	E. or W. Distance.
0·97	0·24	51	49·49	12·34	1	0·97	0·26	51	49·26	13·20
1·94	0·48	52	50·46	12·58	2	1·93	0·52	52	50·23	13·46
2·91	0·72	53	51·43	12·82	3	2·90	0·78	53	51·19	13·72
3·88	0·97	54	52·40	13·06	4	3·86	1·04	54	52·16	13·98
4·85	1·21	55	53·37	13·31	5	4·83	1·29	55	53·13	14·24
5·82	1·45	56	54·34	13·55	6	5·80	1·55	56	54·09	14·49
6·79	1·69	57	55·31	13·79	7	6·76	1·81	57	55·06	14·75
7·76	1·93	58	56·28	14·03	8	7·73	2·07	58	56·02	15·01
8·73	2·18	59	57·25	14·27	9	8·69	2·33	59	56·99	15·27
9·70	2·42	60	58·22	14·52	10	9·66	2·59	60	57·96	15·53
10·67	2·66	61	59·19	14·76	11	10·63	2·85	61	58·92	15·79
11·64	2·90	62	60·16	15·00	12	11·59	3·11	62	59·89	16·05
12·61	3·14	63	61·18	15·24	13	12·56	3·36	63	60·85	16·31
13·58	3·39	64	62·10	15·48	14	13·52	3·62	64	61·82	16·56
14·55	3·63	65	63·07	15·72	15	14·49	3·88	65	62·79	16·82
15·52	3·87	66	64·04	15·97	16	15·45	4·14	66	63·75	17·08
16·50	4·11	67	65·01	16·21	17	16·42	4·40	67	64·72	17·34
17·47	4·35	68	65·98	16·45	18	17·39	4·66	68	65·68	17·60
18·41	4·60	69	66·95	16·69	19	18·35	4·92	69	66·65	17·86
19·41	4·84	70	67·92	16·94	20	19·32	5·18	70	67·61	18·12
20·38	5·08	71	68·89	17·18	21	20·28	5·44	71	68·58	18·38
21·35	5·32	72	69·86	17·42	22	21·25	5·69	72	69·55	18·63
22·32	5·56	73	70·83	17·66	23	22·22	5·95	73	70·51	18·89
23·29	5·81	74	71·80	17·90	24	23·18	6·21	74	71·48	19·15
24·26	6·05	75	72·77	18·14	25	24·15	6·47	75	72·44	19·41
25·23	6·29	76	73·74	18·39	26	25·11	6·73	76	73·41	19·67
26·20	6·53	77	74·71	18·63	27	26·08	6·99	77	74·38	19·93
27·17	6·77	78	75·68	18·87	28	27·05	7·25	78	75·34	20·19
28·14	7·02	79	76·65	19·11	29	28·01	7·51	79	76·31	20·45
29·11	7·26	80	77·62	19·35	30	28·98	7·76	80	77·27	20·71
30·08	7·50	81	78·59	19·60	31	29·94	8·02	81	78·24	20·96
31·05	7·74	82	79·56	19·84	32	30·91	8·28	82	79·21	21·22
32·02	7·98	83	80·53	20·08	33	31·88	8·54	83	80·17	21·48
32·99	8·23	84	81·50	20·32	34	32·84	8·80	84	81·14	21·74
33·96	8·47	85	82·48	20·56	35	33·81	9·06	85	82·10	22·00
34·93	8·71	86	83·45	20·81	36	34·77	9·32	86	83·07	22·26
35·90	8·95	87	84·42	21·05	37	35·74	9·58	87	84·04	22·52
36·87	9·19	88	85·39	21·29	38	36·71	9·84	88	85·00	22·78
37·84	9·43	89	86·36	21·53	39	37·67	10·09	89	85·97	23·03
38·81	9·68	90	87·33	21·77	40	38·64	10·35	90	86·93	23·29
39·78	9·92	91	88·30	22·01	41	39·60	10·61	91	87·90	23·55
40·75	10·16	92	89·27	22·26	42	40·57	10·87	92	88·87	23·81
41·72	10·40	93	90·24	22·50	43	41·53	11·13	93	89·83	24·07
42·69	10·64	94	91·21	22·74	44	42·50	11·39	94	90·80	24·33
43·66	10·89	95	92·18	22·98	45	43·47	11·65	95	91·76	24·59
44·63	11·13	96	93·15	23·22	46	44·43	11·91	96	92·73	24·85
45·60	11·37	97	94·12	23·47	47	45·40	12·16	97	93·69	25·11
46·57	11·61	98	95·09	23·71	48	46·36	12·42	98	94·66	25·36
47·54	11·85	99	96·06	23·95	49	47·33	12·68	99	95·63	25·62
48·51	12·10	100	97·03	24·19	50	48·30	12·94	100	96·60	25·88
E. or W.	N. or S.		E. or W.	N. or S.		E. or W.	N. or S.		E. or W.	N. or S.
76°					75°					

16°						17°					
Bearing Lengths.	N. or S. Distance.	E. or W. Distance.	Bearing Lengths.	N. or S. Distance.	E. or W. Distance.	Bearing Lengths.	N. or S. Distance.	E. or W. Distance.	Bearing Lengths.	N. or S. Distance.	E. or W. Distance.
1	0·96	0·28	51	49·02	14·06	1	0·96	0·29	51	48·77	14·91
2	1·92	0·55	52	49·99	14·33	2	1·91	0·58	52	49·73	15·20
3	2·88	0·83	53	50·95	14·61	3	2·87	0·88	53	50·68	15·50
4	3·85	1·10	54	51·91	14·88	4	3·83	1·17	54	51·64	15·79
5	4·81	1·38	55	52·87	15·16	5	4·78	1·46	55	52·60	16·08
6	5·77	1·65	56	53·83	15·44	6	5·74	1·75	56	53·55	16·37
7	6·73	1·93	57	54·79	15·71	7	6·69	2·05	57	54·51	16·67
8	7·69	2·21	58	55·75	15·99	8	7·65	2·34	58	55·47	16·90
9	8·65	2·48	59	56·71	16·26	9	8·61	2·63	59	56·42	17·25
10	9·61	2·76	60	57·68	16·54	10	9·56	2·92	60	57·38	17·54
11	10·57	3·03	61	58·64	16·81	11	10·52	3·22	61	58·33	17·83
12	11·54	3·31	62	59·60	17·09	12	11·48	3·51	62	59·29	18·13
13	12·50	3·58	63	60·56	17·37	13	12·43	3·80	63	60·25	18·42
14	13·46	3·86	64	61·52	17·64	14	13·39	4·09	64	61·20	18·71
15	14·42	4·13	65	62·48	17·92	15	14·34	4·39	65	62·16	19·00
16	15·38	4·41	66	63·44	18·19	16	15·30	4·68	66	63·12	19·30
17	16·34	4·69	67	64·40	18·47	17	16·26	4·97	67	64·07	19·59
18	17·30	4·96	68	65·37	18·74	18	17·21	5·26	68	65·03	19·88
19	18·26	5·24	69	66·33	19·02	19	18·17	5·56	69	65·98	20·17
20	19·23	5·51	70	67·29	19·29	20	19·13	5·85	70	66·94	20·47
21	20·19	5·79	71	68·25	19·57	21	20·08	6·14	71	67·90	20·76
22	21·15	6·06	72	69·21	19·85	22	21·04	6·43	72	68·85	21·05
23	22·11	6·34	73	70·17	20·12	23	21·99	6·72	73	69·81	21·34
24	23·07	6·62	74	71·13	20·40	24	22·95	7·02	74	70·77	21·64
25	24·03	6·89	75	72·09	20·67	25	23·91	7·31	75	71·72	21·93
26	24·99	7·17	76	73·06	20·95	26	24·86	7·60	76	72·68	22·22
27	25·96	7·44	77	74·02	21·22	27	25·82	7·89	77	73·64	22·51
28	26·92	7·72	78	74·98	21·50	28	26·78	8·19	78	74·60	22·80
29	27·88	7·99	79	75·94	21·78	29	27·73	8·48	79	75·55	23·10
30	28·84	8·27	80	76·90	22·05	30	28·69	8·77	80	76·50	23·39
31	29·80	8·54	81	77·86	22·33	31	29·65	9·06	81	77·46	23·68
32	30·76	8·82	82	78·82	22·60	32	30·60	9·36	82	78·42	23·97
33	31·72	9·10	83	79·78	22·88	33	31·56	9·65	83	79·37	24·27
34	32·68	9·37	84	80·75	23·15	34	32·51	9·94	84	80·33	24·56
35	33·64	9·65	85	81·71	23·43	35	33·47	10·23	85	81·29	24·85
36	34·61	9·92	86	82·67	23·70	36	34·43	10·53	86	82·24	25·14
37	35·57	10·20	87	83·63	23·98	37	35·38	10·82	87	83·20	25·44
38	36·53	10·47	88	84·59	24·26	38	36·34	11·11	88	84·15	25·73
39	37·49	10·75	89	85·55	24·56	39	37·30	11·40	89	85·11	26·02
40	38·45	11·03	90	86·51	24·81	40	38·25	11·69	90	86·07	26·31
41	39·41	11·30	91	87·47	25·08	41	39·21	11·99	91	87·02	26·61
42	40·37	11·58	92	88·44	25·36	42	40·16	12·28	92	87·98	26·90
43	41·33	11·85	93	89·40	25·63	43	41·12	12·57	93	88·94	27·19
44	42·30	12·13	94	90·36	25·91	44	42·08	12·86	94	89·89	27·48
45	43·26	12·40	95	91·32	26·19	45	43·03	13·16	95	90·85	27·78
46	44·22	12·68	96	92·29	26·46	46	43·99	13·45	96	91·81	28·07
47	45·18	12·96	97	93·24	26·74	47	44·95	13·74	97	92·76	28·36
48	46·14	13·23	98	94·20	27·01	48	45·90	14·03	98	93·72	28·65
49	47·10	13·51	99	95·16	27·29	49	46·86	14·33	99	94·67	28·94
50	48·06	13·78	100	96·13	27·56	50	47·82	14·62	100	95·63	29·24
	E. or W.	N. or S.		E. or W.	N. or S.		E. or W.	N. or S.		E. or W.	N. or S.
74°						73°					

18°					
Bearing Lengths.	N. or S. Distance.	E. or W. Distance.	Bearing Lengths.	N. or S. Distance.	E. or W. Distance.
1	0·95	0·31	51	48·50	15·76
2	1·90	0·62	52	49·45	16·07
3	2·85	0·93	53	50·41	16·38
4	3·80	1·24	54	51·36	16·69
5	4·76	1·55	55	52·31	17·00
6	5·71	1·85	56	53·26	17·30
7	6·66	2·16	57	54·21	17·61
8	7·61	2·47	58	55·16	17·92
9	8·56	2·78	59	56·11	18·23
10	9·51	3·09	60	57·06	18·54
11	10·46	3·40	61	58·01	18·85
12	11·41	3·71	62	58·97	19·16
13	12·36	4·02	63	59·92	19·47
14	13·31	4·33	64	60·87	19·78
15	14·27	4·64	65	61·82	20·09
16	15·22	4·94	66	62·77	20·40
17	16·17	5·25	67	63·72	20·70
18	17·12	5·56	68	64·67	21·01
19	18·07	5·87	69	65·62	21·32
20	19·02	6·18	70	66·57	21·63
21	19·97	6·49	71	67·53	21·94
22	20·92	6·80	72	68·48	22·25
23	21·87	7·11	73	69·43	22·56
24	22·83	7·42	74	70·38	22·87
25	23·78	7·73	75	71·33	23·18
26	24·73	8·03	76	72·28	23·49
27	25·68	8·34	77	73·23	23·79
28	26·63	8·65	78	74·18	24·10
29	27·58	8·96	79	75·13	24·41
30	28·53	9·27	80	76·08	24·72
31	29·48	9·58	81	77·04	25·03
32	30·43	9·89	82	77·99	25·34
33	31·38	10·20	83	78·94	25·65
34	32·34	10·51	84	79·89	25·96
35	33·29	10·82	85	80·84	26·27
36	34·24	11·12	86	81·79	26·58
37	35·19	11·43	87	82·74	26·88
38	36·14	11·74	88	83·69	27·19
39	37·09	12·05	89	84·64	27·50
40	38·04	12·36	90	85·60	27·81
41	38·99	12·67	91	86·55	28·12
42	39·94	12·98	92	87·50	28·43
43	40·90	13·29	93	88·45	28·74
44	41·85	13·60	94	89·40	29·05
45	42·80	13·91	95	90·35	29·36
46	43·75	14·21	96	91·30	29·67
47	44·70	14·52	97	92·25	29·97
48	45·65	14·83	98	93·20	30·28
49	46·60	15·14	99	94·15	30·59
50	47·55	15·45	100	95·11	30·90
	E. or W.	N. or S.		E. or W.	N. or S.
72°					

19°					
Bearing Lengths.	N. or S. Distance.	E. or W. Distance.	Bearing Lengths.	N. or S. Distance.	E. or W. Distance.
1	0·95	0·33	51	48·22	16·60
2	1·89	0·65	52	49·17	16·93
3	2·84	0·98	53	50·11	17·26
4	3·78	1·30	54	51·06	17·58
5	4·73	1·63	55	52·00	17·91
6	5·67	1·95	56	52·95	18·23
7	6·62	2·28	57	53·89	18·56
8	7·56	2·60	58	54·84	18·88
9	8·51	2·93	59	55·79	19·21
10	9·46	3·26	60	56·73	19·53
11	10·40	3·58	61	57·68	19·86
12	11·35	3·91	62	58·62	20·19
13	12·29	4·23	63	59·57	20·51
14	13·24	4·56	64	60·51	20·84
15	14·18	4·88	65	61·46	21·16
16	15·13	5·21	66	62·40	21·49
17	16·07	5·53	67	63·35	21·81
18	17·02	5·86	68	64·30	22·14
19	17·96	6·19	69	65·24	22·46
20	18·91	6·51	70	66·19	22·79
21	19·86	6·84	71	67·13	23·12
22	20·80	7·16	72	68·08	23·44
23	21·75	7·49	73	69·02	23·77
24	22·69	7·81	74	69·97	24·09
25	23·64	8·14	75	70·91	24·42
26	24·58	8·46	76	71·86	24·74
27	25·53	8·79	77	72·81	25·07
28	26·47	9·12	78	73·75	25·39
29	27·42	9·44	79	74·70	25·72
30	28·37	9·77	80	75·64	26·05
31	29·31	10·09	81	76·59	26·37
32	30·26	10·42	82	77·53	26·70
33	31·20	10·74	83	78·48	27·02
34	32·15	11·07	84	79·42	27·35
35	33·09	11·39	85	80·37	27·67
36	34·04	11·72	86	81·31	28·00
37	34·98	12·05	87	82·26	28·32
38	35·93	12·37	88	83·21	28·65
39	36·87	12·70	89	84·15	28·98
40	37·82	13·02	90	85·10	29·30
41	38·77	13·35	91	86·04	29·63
42	39·71	13·67	92	86·99	29·95
43	40·66	14·00	93	87·93	30·28
44	41·60	14·32	94	88·88	30·60
45	42·55	14·65	95	89·82	30·93
46	43·49	14·98	96	90·77	31·25
47	44·44	15·30	97	91·72	31·58
48	45·38	15·63	98	92·66	31·91
49	46·33	15·95	99	93·61	32·23
50	47·28	16·28	100	94·55	32·56
	E. or W.	N. or S.		E. or W.	N. or S.
71°					

20°						21°					
Bearing Lengths.	N. or S. Distance.	E. or W. Distance.	Bearing Lengths.	N. or S. Distance.	E. or W. Distance.	Bearing Lengths.	N. or S. Distance.	E. or W. Distance.	Bearing Lengths.	N. or S. Distance.	E. or W. Distance.
1	0·94	0·34	51	47·92	17·44	1	0·93	0·36	51	47·61	18·28
2	1·88	0·68	52	48·86	17·79	2	1·87	0·72	52	48·55	18·63
3	2·82	1·03	53	49·80	18·13	3	2·80	1·08	53	49·48	18·99
4	3·76	1·37	54	50·74	18·47	4	3·73	1·43	54	50·41	19·35
5	4·70	1·71	55	51·68	18·81	5	4·67	1·79	55	51·35	19·71
6	5·64	2·05	56	52·62	19·15	6	5·60	2·15	56	52·28	20·07
7	6·58	2·39	57	53·56	19·50	7	6·54	2·51	57	53·21	20·43
8	7·52	2·74	58	54·50	19·84	8	7·47	2·87	58	54·15	20·78
9	8·46	3·08	59	55·44	20·18	9	8·40	3·23	59	55·08	21·14
10	9·40	3·42	60	56·38	20·52	10	9·34	3·58	60	56·01	21·50
11	10·34	3·76	61	57·32	20·86	11	10·27	3·94	61	56·95	21·86
12	11·28	4·10	62	58·26	21·21	12	11·20	4·30	62	57·88	22·22
13	12·22	4·45	63	59·20	21·55	13	12·14	4·66	63	58·82	22·58
14	13·16	4·79	64	60·14	21·89	14	13·07	5·02	64	59·75	22·93
15	14·10	5·13	65	61·08	22·23	15	14·00	5·38	65	60·68	23·29
16	15·04	5·47	66	62·02	22·57	16	14·94	5·73	66	61·62	23·65
17	15·97	5·81	67	62·96	22·92	17	15·87	6·09	67	62·55	24·01
18	16·91	6·16	68	63·90	23·26	18	16·80	6·45	68	63·48	24·37
19	17·85	6·50	69	64·84	23·60	19	17·74	6·81	69	64·42	24·73
20	18·79	6·84	70	65·78	23·94	20	18·67	7·17	70	65·35	25·08
21	19·73	7·18	71	66·72	24·28	21	19·61	7·53	71	66·28	25·44
22	20·67	7·52	72	67·66	24·63	22	20·54	7·88	72	67·22	25·80
23	21·61	7·87	73	68·60	24·97	23	21·47	8·24	73	68·15	26·16
24	22·55	8·21	74	69·54	25·31	24	22·41	8·60	74	69·09	26·52
25	23·49	8·55	75	70·48	25·65	25	23·34	8·96	75	70·02	26·88
26	24·43	8·89	76	71·42	25·99	26	24·27	9·32	76	70·95	27·23
27	25·37	9·23	77	72·36	26·34	27	25·21	9·68	77	71·89	27·59
28	26·31	9·58	78	73·30	26·68	28	26·14	10·03	78	72·82	27·95
29	27·25	9·92	79	74·24	27·02	29	27·07	10·39	79	73·75	28·31
30	28·19	10·26	80	75·18	27·36	30	28·01	10·75	80	74·68	28·67
31	29·13	10·60	81	76·12	27·70	31	28·94	11·11	81	75·62	29·03
32	30·07	10·94	82	77·06	28·05	32	29·87	11·47	82	76·55	29·39
33	31·01	11·29	83	77·99	28·39	33	30·81	11·83	83	77·49	29·74
34	31·95	11·63	84	78·93	28·73	34	31·74	12·18	84	78·42	30·10
35	32·89	11·97	85	79·87	29·07	35	32·68	12·54	85	79·35	30·46
36	33·83	12·31	86	80·81	29·41	36	33·61	12·90	86	80·29	30·82
37	34·77	12·65	87	81·75	29·76	37	34·54	13·26	87	81·22	31·18
38	35·71	13·00	88	82·69	30·10	38	35·48	13·62	88	82·16	31·54
39	36·65	13·34	89	83·63	30·44	39	36·41	13·98	89	83·09	31·89
40	37·59	13·68	90	84·57	30·78	40	37·34	14·33	90	84·02	32·25
41	38·53	14·02	91	85·51	31·12	41	38·28	14·69	91	84·96	32·61
42	39·47	14·36	92	86·45	31·47	42	39·21	15·05	92	85·89	32·97
43	40·41	14·71	93	87·39	31·81	43	40·14	15·41	93	86·82	33·33
44	41·35	15·05	94	88·33	32·15	44	41·08	15·77	94	87·76	33·69
45	42·29	15·39	95	89·27	32·49	45	42·01	16·13	95	88·69	34·04
46	43·23	15·73	96	90·21	32·83	46	42·94	16·48	96	89·62	34·40
47	44·17	16·07	97	91·15	33·18	47	43·88	16·84	97	90·56	34·76
48	45·11	16·42	98	92·09	33·52	48	44·81	17·20	98	91·49	35·12
49	46·04	16·76	99	93·03	33·86	49	45·75	17·56	99	92·42	35·48
50	46·98	17·10	100	93·97	34·20	50	46·68	17·92	100	93·36	35·84
	E. or W.	N. or S.		E. or W.	N. or S.		E. or W.	N. or S.		E. or W.	N. or S.
70°						69°					

22°						23°					
Bearing Lengths.	N. or S. Distance.	E. or W. Distance.	Bearing Lengths.	N. or S. Distance.	E. or W. Distance.	Bearing Lengths.	N. or S. Distance.	E. or W. Distance.	Bearing Lengths.	N. or S. Distance.	E. or W. Distance.
1	0·93	0·37	51	47·29	19·10	1	0·92	0·39	51	46·95	19·93
2	1·85	0·75	52	48·21	19·48	2	1·84	0·78	52	47·87	20·32
3	2·78	1·12	53	49·14	19·85	3	2·76	1·17	53	48·79	20·17
4	3·71	1·50	54	50·07	20·23	4	3·68	1·56	54	49·71	21·10
5	4·64	1·87	55	51·00	20·60	5	4·60	1·95	55	50·63	21·49
6	5·56	2·25	56	51·92	20·98	6	5·52	2·34	56	51·55	21·88
7	6·49	2·62	57	52·85	21·35	7	6·44	2·74	57	52·47	22·27
8	7·42	3·00	58	53·78	21·73	8	7·36	3·13	58	53·39	22·66
9	8·34	3·37	59	54·70	22·10	9	8·28	3·52	59	54·31	23·05
10	9·27	3·75	60	55·63	22·48	10	9·21	3·91	60	55·23	23·44
11	10·20	4·12	61	56·56	22·85	11	10·13	4·30	61	56·15	23·83
12	11·13	4·50	62	57·49	23·23	12	11·05	4·69	62	57·07	24·23
13	12·05	4·87	63	58·41	23·60	13	11·97	5·08	63	57·99	24·62
14	12·98	5·24	64	59·34	23·97	14	12·89	5·47	64	58·91	25·01
15	13·91	5·62	65	60·27	24·35	15	13·81	5·86	65	59·83	25·40
16	14·83	5·99	66	61·19	24·72	16	14·73	6·25	66	60·75	25·79
17	15·76	6·37	67	62·12	25·10	17	15·65	6·64	67	61·67	26·18
18	16·69	6·74	68	63·05	25·47	18	16·57	7·03	68	62·59	26·57
19	17·62	7·12	69	63·98	25·85	19	17·49	7·42	69	63·51	26·96
20	18·54	7·49	70	64·90	26·22	20	18·41	7·81	70	64·44	27·35
21	19·47	7·87	71	65·83	26·60	21	19·33	8·21	71	65·36	27·74
22	20·40	8·24	72	66·76	26·97	22	20·25	8·60	72	66·28	28·13
23	21·33	8·62	73	67·68	27·35	23	21·17	8·99	73	67·20	28·52
24	22·25	8·99	74	68·61	27·72	24	22·09	9·38	74	68·12	28·91
25	23·18	9·37	75	69·54	28·10	25	23·01	9·77	75	69·04	29·30
26	24·11	9·74	76	70·47	28·47	26	23·93	10·16	76	69·96	29·70
27	25·03	10·11	77	71·39	28·84	27	24·85	10·55	77	70·88	30·09
28	25·96	10·49	78	72·32	29·22	28	25·77	10·94	78	71·80	30·48
29	26·89	10·86	79	73·25	29·59	29	26·69	11·33	79	72·72	30·87
30	27·82	11·24	80	74·17	29·97	30	27·62	11·72	80	73·64	31·26
31	28·74	11·61	81	75·10	30·34	31	28·54	12·11	81	74·56	31·65
32	29·67	11·99	82	76·03	30·72	32	29·46	12·50	82	75·48	32·04
33	30·60	12·36	83	76·96	31·09	33	30·38	12·89	83	76·40	32·43
34	31·52	12·74	84	77·88	31·47	34	31·30	13·28	84	77·32	32·82
35	32·45	13·11	85	78·81	31·84	35	32·22	13·68	85	78·24	33·21
36	33·38	13·49	86	79·74	32·22	36	33·14	14·07	86	79·16	33·60
37	34·31	13·86	87	80·66	32·59	37	34·06	14·46	87	80·08	33·99
38	35·23	14·24	88	81·59	32·97	38	34·98	14·85	88	81·00	34·38
39	36·16	14·61	89	82·52	33·34	39	35·90	15·24	89	81·92	34·78
40	37·09	14·98	90	83·45	33·71	40	36·82	15·63	90	82·85	35·17
41	38·01	15·36	91	84·37	34·09	41	37·74	16·02	91	83·77	35·56
42	38·94	15·73	92	85·30	34·46	42	38·66	16·41	92	84·69	35·95
43	39·87	16·11	93	86·23	34·84	43	39·58	16·80	93	85·61	36·34
44	40·80	16·48	94	87·16	35·21	44	40·50	17·19	94	86·53	36·73
45	41·72	16·86	95	88·08	35·59	45	41·42	17·58	95	87·45	37·12
46	42·65	17·23	96	89·01	35·96	46	42·34	17·97	96	88·37	37·51
47	43·58	17·61	97	89·94	36·34	47	43·26	18·36	97	89·29	37·90
48	44·50	17·98	98	90·86	36·71	48	44·18	18·76	98	90·21	38·29
49	45·43	18·36	99	91·79	37·09	49	45·10	19·15	99	91·13	38·68
50	46·36	18·73	100	92·72	37·46	50	46·03	19·54	100	92·05	39·07
	E. or W.	N. or S.		E. or W.	N. or S.		E. or W.	N. or S.		E. or W.	N. or S.
68°						67°					

24°						25°					
Bearing Lengths.	N. or S. Distance.	E. or W. Distance.	Bearing Lengths.	N. or S. Distance.	E. or W. Distance.	Bearing Lengths.	N. or S. Distance.	E. or W. Distance.	Bearing Lengths.	N. or S. Distance.	E. or W. Distance.
1	0·91	0·41	51	46·59	20·74	1	0·91	0·42	51	46·22	21·55
2	1·83	0·81	52	47·50	21·15	2	1·81	0·85	52	47·13	21·98
3	2·74	1·22	53	48·42	21·56	3	2·72	1·27	53	48·04	22·40
4	3·65	1·63	54	49·33	21·96	4	3·63	1·69	54	48·94	22·82
5	4·57	2·03	55	50·25	22·37	5	4·53	2·11	55	49·85	23·24
6	5·48	2·44	56	51·16	22·78	6	5·44	2·54	56	50·75	23·67
7	6·39	2·85	57	52·07	23·18	7	6·34	2·96	57	51·66	24·09
8	7·31	3·25	58	52·99	23·59	8	7·25	3·38	58	52·57	24·51
9	8·22	3·66	59	53·90	24·00	9	8·16	3·80	59	53·47	24·93
10	9·14	4·07	60	54·81	24·40	10	9·06	4·23	60	54·38	25·36
11	10·05	4·47	61	55·73	24·81	11	9·97	4·65	61	55·28	25·78
12	10·96	4·88	62	56·64	25·22	12	10·88	5·07	62	56·19	26·20
13	11·88	5·29	63	57·55	25·62	13	11·78	5·49	63	57·10	26·62
14	12·79	5·69	64	58·47	26·03	14	12·69	5·92	64	58·00	27·05
15	13·70	6·10	65	59·38	26·44	15	13·59	6·34	65	58·91	27·47
16	14·62	6·51	66	60·29	26·84	16	14·50	6·76	66	59·82	27·89
17	15·53	6·91	67	61·21	27·25	17	15·41	7·18	67	60·72	28·32
18	16·44	7·32	68	62·12	27·66	18	16·31	7·61	68	61·63	28·74
19	17·36	7·73	69	63·03	28·06	19	17·22	8·03	69	62·54	29·16
20	18·27	8·13	70	63·95	28·47	20	18·13	8·45	70	63·44	29·60
21	19·18	8·54	71	64·86	28·88	21	19·03	8·87	71	64·35	30·01
22	20·10	8·95	72	65·78	29·28	22	19·94	9·30	72	65·25	30·43
23	21·01	9·35	73	66·69	29·69	23	20·85	9·72	73	66·16	30·85
24	21·93	9·76	74	67·60	30·10	24	21·75	10·14	74	67·07	31·27
25	22·84	10·17	75	68·52	30·50	25	22·66	10·57	75	67·97	31·70
26	23·75	10·58	76	69·43	30·91	26	23·56	10·99	76	68·88	32·12
27	24·67	10·98	77	70·34	31·32	27	24·47	11·41	77	69·79	32·54
28	25·58	11·39	78	71·26	31·72	28	25·38	11·83	78	70·69	32·96
29	26·49	11·80	79	72·17	32·13	29	26·28	12·26	79	71·60	33·39
30	27·41	12·20	80	73·08	32·54	30	27·19	12·68	80	72·50	33·81
31	28·32	12·61	81	74·00	32·94	31	28·10	13·10	81	73·41	34·23
32	29·23	13·02	82	74·91	33·35	32	29·00	13·52	82	74·32	34·65
33	30·15	13·42	83	75·82	33·76	33	29·91	13·95	83	75·22	35·08
34	31·06	13·83	84	76·74	34·16	34	30·81	14·37	84	76·13	35·50
35	31·97	14·24	85	77·65	34·57	35	31·72	14·79	85	77·04	35·92
36	32·89	14·64	86	78·56	34·98	36	32·63	15·21	86	77·94	36·35
37	33·80	15·05	87	79·48	35·38	37	33·53	15·64	87	78·85	36·77
38	34·71	15·46	88	80·39	35·79	38	34·44	16·06	88	79·76	37·19
39	35·63	15·86	89	81·31	36·20	39	35·35	16·48	89	80·66	37·61
40	36·54	16·27	90	82·22	36·60	40	36·25	16·90	90	81·57	38·04
41	37·46	16·68	91	83·13	37·01	41	37·16	17·33	91	82·47	38·46
42	38·37	17·08	92	84·05	37·42	42	38·06	17·75	92	83·38	38·88
43	39·28	17·49	93	84·96	37·82	43	38·97	18·17	93	84·29	39·30
44	40·20	17·90	94	85·87	38·23	44	39·88	18·60	94	85·19	39·73
45	41·11	18·30	95	86·79	38·64	45	40·78	19·02	95	86·10	40·15
46	42·02	18·71	96	87·70	39·04	46	41·69	19·44	96	87·01	40·57
47	42·94	19·12	97	88·61	39·45	47	42·60	19·86	97	87·91	40·99
48	43·85	19·52	98	89·53	39·86	48	43·50	20·29	98	88·82	41·42
49	44·76	19·93	99	90·44	40·26	49	44·41	20·71	99	89·72	41·84
50	45·68	20·34	100	91·35	40·67	50	45·32	21·13	100	90·63	42·26
	E. or W.	N. or S.		E. or W.	N. or S.		E. or W.	N. or S.		E. or W.	N. or S.
66°						65°					

26°						27°					
Bearing Lengths.	N. or S. Distance.	E. or W. Distance.	Bearing Lengths.	N. or S. Distance.	E. or W. Distance.	Bearing Lengths.	N. or S. Distance.	E. or W. Distance.	Bearing Lengths.	N. or S. Distance.	E. or W. Distance.
1	0·90	0·44	51	45·84	22·36	1	0·89	0·45	51	45·44	23·15
2	1·80	0·88	52	46·74	22·80	2	1·78	0·91	52	46·33	23·61
3	2·70	1·32	53	47·64	23·23	3	2·67	1·36	53	47·22	24·06
4	3·60	1·75	54	48·53	23·67	4	3·56	1·82	54	48·11	24·52
5	4·49	2·19	55	49·43	24·11	5	4·46	2·27	55	49·01	24·97
6	5·39	2·63	56	50·33	24·55	6	5·35	2·72	56	49·90	25·42
7	6·29	3·07	57	51·23	24·99	7	6·24	3·18	57	50·79	25·88
8	7·19	3·50	58	52·13	25·43	8	7·13	3·63	58	51·68	26·33
9	8·09	3·95	59	53·03	25·86	9	8·02	4·09	59	52·57	26·79
10	8·99	4·38	60	53·93	26·30	10	8·91	4·54	60	53·46	27·24
11	9·89	4·82	61	54·83	26·74	11	9·80	4·99	61	54·35	27·69
12	10·79	5·26	62	55·73	27·18	12	10·69	5·45	62	55·24	28·15
13	11·68	5·70	63	56·62	27·62	13	11·58	5·90	63	56·13	28·60
14	12·58	6·14	64	57·52	28·06	14	12·47	6·36	64	57·02	29·06
15	13·48	6·58	65	58·42	28·49	15	13·37	6·81	65	57·92	29·51
16	14·38	7·01	66	59·32	28·93	16	14·26	7·26	66	58·81	29·96
17	15·28	7·45	67	60·22	29·37	17	15·15	7·72	67	59·70	30·42
18	16·18	7·89	68	61·12	29·81	18	16·04	8·17	68	60·59	30·87
19	17·08	8·33	69	62·02	30·25	19	16·93	8·63	69	61·48	31·33
20	17·98	8·77	70	62·92	30·69	20	17·82	9·08	70	62·37	31·78
21	18·87	9·21	71	63·81	31·12	21	18·71	9·53	71	63·26	32·23
22	19·77	9·64	72	64·71	31·56	22	19·60	9·99	72	64·15	32·69
23	20·67	10·08	73	65·61	32·00	23	20·49	10·44	73	65·04	33·14
24	21·57	10·52	74	66·51	32·44	24	21·38	10·90	74	65·93	33·60
25	22·47	10·96	75	67·41	32·88	25	22·28	11·35	75	66·83	34·05
26	23·37	11·40	76	68·31	33·32	26	23·17	11·80	76	67·72	34·50
27	24·27	11·84	77	69·21	33·75	27	24·06	12·26	77	68·61	34·96
28	25·17	12·27	78	70·11	34·19	28	24·95	12·71	78	69·50	35·41
29	26·06	12·71	79	71·00	34·63	29	25·84	13·17	79	70·39	35·87
30	26·96	13·15	80	71·90	35·07	30	26·73	13·62	80	71·28	36·32
31	27·86	13·59	81	72·80	35·51	31	27·63	14·07	81	72·17	36·77
32	28·76	14·03	82	73·70	35·95	32	28·51	14·53	82	73·06	37·23
33	29·66	14·47	83	74·60	36·38	33	29·40	14·98	83	73·95	37·68
34	30·56	14·90	84	75·50	36·82	34	30·29	15·44	84	74·84	38·14
35	31·46	15·34	85	76·40	37·26	35	31·19	15·89	85	75·74	38·59
36	32·36	15·78	86	77·30	37·70	36	32·08	16·34	86	76·63	39·04
37	33·26	16·22	87	78·19	38·14	37	32·97	16·80	87	77·52	39·50
38	34·15	16·66	88	79·09	38·58	38	33·86	17·25	88	78·41	39·95
39	35·05	17·10	89	79·99	39·02	39	34·75	17·71	89	79·30	40·41
40	35·95	17·53	90	80·89	39·45	40	35·64	18·16	90	80·19	40·86
41	36·85	17·97	91	81·79	39·89	41	36·53	18·61	91	81·08	41·31
42	37·75	18·41	92	82·69	40·33	42	37·42	19·07	92	81·97	41·77
43	38·65	18·85	93	83·59	40·77	43	38·31	19·52	93	82·86	42·22
44	39·55	19·29	94	84·49	41·21	44	39·20	19·98	94	83·75	42·68
45	40·45	19·73	95	85·39	41·65	45	40·10	20·43	95	84·65	43·13
46	41·34	20·17	96	86·28	42·08	46	40·99	20·88	96	85·54	43·58
47	42·24	20·60	97	87·18	42·52	47	41·88	21·34	97	86·43	44·04
48	43·14	21·04	98	88·08	42·96	48	42·77	21·79	98	87·32	44·49
49	44·04	21·48	99	88·98	43·40	49	43·66	22·25	99	88·21	44·95
50	44·94	21·92	100	89·88	43·84	50	44·55	22·70	100	89·10	45·40
	E. or W.	N. or S.		E. or W.	N. or S.		E. or W.	N. or S.		E. or W.	N. or S.
64°						63°					

28°						29°					
Bearing Lengths.	N. or S. Distance.	E. or W. Distance.	Bearing Lengths.	N. or S. Distance.	E. or W. Distance.	Bearing Lengths.	N. or S. Distance.	E. or W. Distance.	Bearing Lengths.	N. or S. Distance.	E. or W. Distance.
1	0·88	0·47	51	45·03	23·94	1	0·87	0·48	51	44·61	24·73
2	1·77	0·94	52	45·91	24·41	2	1·75	0·97	52	45·48	25·21
3	2·65	1·41	53	46·80	24·88	3	2·62	1·45	53	46·35	25·69
4	3·53	1·88	54	47·68	25·35	4	3·50	1·94	54	47·23	26·18
5	4·41	2·35	55	48·56	25·82	5	4·37	2·42	55	48·10	26·66
6	5·30	2·82	56	49·45	26·29	6	5·25	2·91	56	48·98	27·15
7	6·18	3·29	57	50·33	26·76	7	6·12	3·39	57	49·85	27·63
8	7·06	3·76	58	51·21	27·23	8	7·00	3·88	58	50·73	28·12
9	7·95	4·23	59	52·09	27·70	9	7·87	4·36	59	51·60	28·60
10	8·83	4·69	60	52·98	28·17	10	8·75	4·85	60	52·48	29·09
11	9·71	5·16	61	53·86	28·64	11	9·62	5·33	61	53·35	29·57
12	10·60	5·63	62	54·74	29·11	12	10·50	5·82	62	54·23	30·06
13	11·48	6·10	63	55·63	29·58	13	11·37	6·30	63	55·10	30·54
14	12·36	6·57	64	56·51	30·05	14	12·24	6·79	64	55·98	31·03
15	13·24	7·04	65	57·39	30·52	15	13·12	7·27	65	56·85	31·51
16	14·13	7·51	66	58·27	30·99	16	13·99	7·76	66	57·72	32·00
17	15·01	7·98	67	59·16	31·45	17	14·87	8·24	67	58·60	32·48
18	15·89	8·15	68	60·04	31·92	18	15·74	8·73	68	59·47	32·97
19	16·78	8·92	69	60·92	32·39	19	16·62	9·21	69	60·35	33·45
20	17·66	9·39	70	61·81	32·83	20	17·49	9·70	70	61·22	33·94
21	18·54	9·86	71	62·69	33·33	21	18·37	10·18	71	62·10	34·42
22	19·42	10·33	72	63·57	33·80	22	19·24	10·67	72	62·97	34·91
23	20·31	10·80	73	64·46	34·27	23	20·12	11·15	73	63·85	35·39
24	21·19	11·27	74	65·34	34·74	24	20·99	11·64	74	64·72	35·88
25	22·07	11·74	75	66·22	35·21	25	21·87	12·12	75	65·60	36·36
26	22·96	12·21	76	67·10	35·68	26	22·74	12·61	76	66·47	36·85
27	23·84	12·68	77	67·99	36·15	27	23·61	13·00	77	67·35	37·33
28	24·72	13·15	78	68·87	36·62	28	24·40	13·57	78	68·22	37·82
29	25·61	13·61	79	69·75	37·09	29	25·36	14·06	79	69·09	38·30
30	26·49	14·08	80	70·64	37·56	30	26·24	14·54	80	69·97	38·78
31	27·37	14·55	81	71·52	38·03	31	27·11	15·03	81	70·84	39·27
32	28·25	15·02	82	72·40	38·50	32	27·99	15·51	82	71·72	39·75
33	29·14	15·49	83	73·28	38·97	33	28·86	16·00	83	72·59	40·24
34	30·02	15·96	84	74·17	39·44	34	29·74	16·48	84	73·47	40·72
35	30·90	16·43	85	75·05	39·91	35	30·61	16·97	85	74·34	41·21
36	31·79	16·90	86	75·93	40·37	36	31·49	17·45	86	75·22	41·69
37	32·67	17·37	87	76·82	40·84	37	32·36	17·94	87	76·09	42·18
38	33·55	17·84	88	77·70	41·31	38	33·23	18·42	88	76·97	42·66
39	34·43	18·31	89	78·58	41·78	39	34·11	18·91	89	77·84	43·15
40	35·32	18·78	90	79·47	42·25	40	34·98	19·39	90	78·72	43·63
41	36·20	19·25	91	80·35	42·72	41	35·86	19·88	91	79·59	44·12
42	37·08	19·72	92	81·23	43·19	42	36·73	20·36	92	80·47	44·60
43	37·97	20·19	93	82·11	43·66	43	37·61	20·85	93	81·34	45·09
44	38·85	20·66	94	83·00	44·13	44	38·48	21·33	94	82·21	45·57
45	39·73	21·13	95	83·88	44·60	45	39·36	21·82	95	83·09	46·06
46	40·62	21·60	96	84·76	45·07	46	40·24	22·30	96	83·96	46·54
47	41·50	22·07	97	85·65	45·54	47	41·11	22·79	97	84·84	47·03
48	42·38	22·53	98	86·53	46·01	48	41·98	23·27	98	85·71	47·51
49	43·26	23·00	99	87·41	46·48	49	42·86	23·76	99	86·59	48·00
50	44·15	23·47	100	88·29	46·95	50	43·73	24·24	100	87·46	48·48
	E. or W.	N. or S.		E. or W.	N. or S.		E. or W.	N. or S.		E. or W.	N. or S.
62°						61°					

30°						31°					
Bearing Lengths.	N. or S. Distance.	E. or W. Distance.	Bearing Lengths.	N. or S. Distance.	E. or W. Distance.	Bearing Lengths.	N. or S. Distance.	E. or W. Distance.	Bearing Lengths.	N. or S. Distance.	E. or W. Distance.
1	0·87	0·50	51	44·17	25·50	1	0·86	0·52	51	43·72	26·27
2	1·73	1·00	52	45·03	26·00	2	1·71	1·03	52	44·57	26·78
3	2·60	1·50	53	45·90	26·50	3	2·57	1·55	53	45·43	27·30
4	3·46	2·00	54	46·77	27·00	4	3·43	2·06	54	46·29	27·81
5	4·33	2·50	55	47·63	27·50	5	4·29	2·58	55	47·14	28·33
6	5·20	3·00	56	48·50	28·00	6	5·14	3·09	56	48·00	28·84
7	6·06	3·50	57	49·36	28·50	7	6·00	3·61	57	48·86	29·36
8	6·93	4·00	58	50·23	29·00	8	6·86	4·12	58	49·72	29·87
9	7·79	4·50	59	51·10	29·50	9	7·71	4·64	59	50·57	30·39
10	8·66	5·00	60	51·96	30·00	10	8·57	5·15	60	51·43	30·90
11	9·53	5·50	61	52·83	30·50	11	9·43	5·67	61	52·29	31·42
12	10·39	6·00	62	53·69	31·00	12	10·29	6·18	62	53·14	31·93
13	11·26	6·50	63	54·56	31·50	13	11·14	6·70	63	54·00	32·45
14	12·12	7·00	64	55·43	32·00	14	12·00	7·21	64	54·86	32·96
15	12·99	7·50	65	56·29	32·50	15	12·86	7·73	65	55·72	33·48
16	13·86	8·00	66	57·16	33·00	16	13·71	8·24	66	56·57	33·99
17	14·72	8·50	67	58·02	33·50	17	14·57	8·76	67	57·43	34·51
18	15·59	9·00	68	58·89	34·00	18	15·43	9·27	68	58·29	35·02
19	16·45	9·50	69	59·76	34·50	19	16·29	9·79	69	59·14	35·54
20	17·32	10·00	70	60·62	35·00	20	17·14	10·30	70	60·00	36·05
21	18·19	10·50	71	61·49	35·50	21	18·00	10·82	71	60·86	36·57
22	19·05	11·00	72	62·35	36·00	22	18·86	11·33	72	61·72	37·08
23	19·92	11·50	73	63·22	36·50	23	19·71	11·85	73	62·57	37·60
24	20·78	12·00	74	64·09	37·00	24	20·57	12·36	74	63·43	38·11
25	21·65	12·50	75	64·95	37·50	25	21·43	12·88	75	64·29	38·63
26	22·52	13·00	76	65·82	38·00	26	22·29	13·39	76	65·14	39·14
27	23·38	13·50	77	66·68	38·50	27	23·14	13·91	77	66·00	39·66
28	24·25	14·00	78	67·55	39·00	28	24·00	14·42	78	66·86	40·17
29	25·11	14·50	79	68·42	39·50	29	24·86	14·94	79	67·72	40·69
30	25·98	15·00	80	69·28	40·00	30	25·72	15·45	80	68·57	41·20
31	26·85	15·50	81	70·15	40·50	31	26·57	15·97	81	69·43	41·72
32	27·71	16·00	82	71·01	41·00	32	27·43	16·48	82	70·29	42·23
33	28·58	16·50	83	71·88	41·50	33	28·29	17·00	83	71·15	42·75
34	29·44	17·00	84	72·75	42·00	34	29·14	17·51	84	72·00	43·26
35	30·31	17·50	85	73·61	42·50	35	30·00	18·03	85	72·86	43·78
36	31·18	18·00	86	74·48	43·00	36	30·86	18·54	86	73·72	44·29
37	32·04	18·50	87	75·35	43·50	37	31·72	19·06	87	74·57	44·81
38	32·91	19·00	88	76·21	44·00	38	32·57	19·57	88	75·43	45·32
39	33·78	19·50	89	77·08	44·50	39	33·43	20·09	89	76·29	45·84
40	34·64	20·00	90	77·94	45·00	40	34·29	20·60	90	77·15	46·35
41	35·51	20·50	91	78·81	45·50	41	35·14	21·12	91	78·00	46·87
42	36·37	21·00	92	79·68	46·00	42	36·00	21·63	92	78·86	47·38
43	37·24	21·50	93	80·54	46·50	43	36·86	22·15	93	79·72	47·00
44	38·11	22·00	94	81·41	47·00	44	37·72	22·66	94	80·57	48·41
45	38·97	22·50	95	82·27	47·50	45	38·57	23·18	95	81·43	48·93
46	39·84	23·00	96	83·14	48·00	46	39·43	23·69	96	82·29	49·44
47	40·70	25·50	97	84·00	48·50	47	40·29	24·21	97	83·15	49·96
48	41·57	24·00	98	84·87	49·00	48	41·14	24·72	98	84·00	50·47
49	42·44	24·50	99	85·74	49·50	49	42·00	25·24	99	84·86	50·99
50	43·30	25·00	100	86·60	50·00	50	42·86	25·75	100	85·72	51·50
	E. or W.	N. or S.		E. or W.	N. or S.		E. or W.	N. or S.		E. or W.	N. or S.
60°						59°					

32°						33°					
Bearing Lengths.	N. or S. Distance.	E. or W. Distance.	Bearing Lengths.	N. or S. Distance.	E. or W. Distance.	Bearing Lengths.	N. or S. Distance.	E. or W. Distance.	Bearing Lengths.	N. or S. Distance.	E. or W. Distance.
1	0·85	0·53	51	43·25	27·03	1	0·84	0·54	51	42·77	27·78
2	1·70	1·06	52	44·10	27·56	2	1·68	1·09	52	43·61	28·32
3	2·54	1·59	53	44·59	28·09	3	2·52	1·63	53	44·45	28·87
4	3·39	2·12	54	45·79	28·62	4	3·35	2·18	54	45·29	29·41
5	4·24	2·65	55	46·64	29·15	5	4·19	2·72	55	46·13	29·96
6	5·09	3·18	56	47·49	29·68	6	5·03	3·27	56	46·97	30·50
7	5·94	3·71	57	48·34	30·21	7	5·87	3·81	57	47·80	31·05
8	6·78	4·23	58	49·19	30·74	8	6·71	4·36	58	48·64	31·59
9	7·63	4·77	59	50·03	31·27	9	7·55	4·90	59	49·48	32·15
10	8·48	5·30	60	50·88	31·80	10	8·39	5·45	60	50·32	32·68
11	9·33	5·83	61	51·73	32·33	11	9·23	5·99	61	51·16	33·22
12	10·18	6·36	62	52·58	32·86	12	10·06	6·54	62	52·00	33·77
13	11·02	6·89	63	53·43	33·38	13	10·90	7·08	63	52·84	34·31
14	11·87	7·42	64	54·28	33·91	14	11·74	7·62	64	53·67	34·86
15	12·72	7·95	65	55·12	34·44	15	12·58	8·17	65	54·51	35·40
16	13·57	8·48	66	55·97	34·97	16	13·42	8·71	66	55·35	35·95
17	14·42	9·01	67	56·82	35·50	17	14·26	9·26	67	56·19	36·49
18	15·26	9·54	68	57·67	36·03	18	15·10	9·80	68	57·03	37·04
19	16·11	10·07	69	58·57	36·56	19	15·93	10·35	69	57·87	37·58
20	16·96	10·60	70	59·36	37·09	20	16·77	10·89	70	58·71	38·13
21	17·81	11·13	71	60·21	37·62	21	17·61	11·44	71	59·55	38·67
22	18·66	11·66	72	61·06	38·15	22	18·45	11·98	72	60·38	39·21
23	19·51	12·19	73	61·91	38·68	23	19·29	12·53	73	61·22	39·76
24	20·35	12·72	74	62·76	39·21	24	20·13	13·07	74	62·06	40·30
25	21·20	13·25	75	63·60	39·74	25	20·97	13·62	75	62·90	40·85
26	22·05	13·78	76	64·45	40·27	26	21·81	14·16	76	63·74	41·39
27	22·90	14·31	77	65·30	40·80	27	22·64	14·71	77	64·58	41·94
28	23·75	14·84	78	66·15	41·33	28	23·48	15·25	78	65·42	42·48
29	24·59	15·37	79	67·00	41·86	29	24·32	15·79	79	66·25	43·03
30	25·44	15·90	80	67·84	42·39	30	25·16	16·34	80	67·09	43·57
31	26·29	16·43	81	68·69	42·92	31	26·00	16·88	81	67·93	44·12
32	27·14	16·96	82	69·54	43·45	32	26·84	17·43	82	68·77	44·66
33	27·99	17·49	83	70·39	43·98	33	27·68	17·97	83	69·61	45·21
34	28·83	18·02	84	71·24	44·51	34	28·51	18·52	84	70·45	45·75
35	29·68	18·55	85	72·08	45·04	35	29·35	19·06	85	71·29	46·29
36	30·53	19·08	86	72·93	45·57	36	30·19	19·61	86	72·13	46·84
37	31·38	19·61	87	73·78	46·10	37	31·03	20·15	87	72·96	47·38
38	32·23	20·14	88	74·63	46·63	38	31·87	20·70	88	73·80	47·93
39	33·07	20·67	89	75·48	47·16	39	32·71	21·24	89	74·64	48·47
40	33·92	21·20	90	76·32	47·69	40	33·55	21·79	90	75·48	49·02
41	34·77	21·73	91	77·17	48·22	41	34·39	22·33	91	76·32	49·56
42	35·62	22·26	92	78·02	48·75	42	35·22	22·88	92	77·16	50·11
43	36·47	22·79	93	78·87	49·28	43	36·06	23·42	93	78·00	50·65
44	37·31	23·32	94	79·72	49·81	44	36·90	23·97	94	78·83	51·20
45	38·16	23·85	95	80·56	50·34	45	37·74	24·51	95	79·67	51·74
46	39·01	24·38	96	81·41	50·87	46	38·58	25·05	96	80·51	52·29
47	39·86	24·91	97	82·26	51·40	47	39·42	25·60	97	81·35	52·83
48	40·71	25·44	98	83·11	51·93	48	40·26	26·14	98	82·19	53·37
49	41·55	25·97	99	83·96	52·46	49	41·09	26·69	99	83·03	53·92
50	42·40	26·50	100	84·81	52·99	50	41·93	27·23	100	83·87	54·46
	E. or W.	N. or S.		E. or W.	N. or S.		E. or W.	N. or S.		E. or W.	N. or S.
58°						57°					

34°						35°					
Bearing Lengths.	N. or S. Distance.	E. or W. Distance.	Bearing Lengths.	N. or S. Distance.	E. or W. Distance.	Bearing Lengths.	N. or S. Distance.	E. or W. Distance.	Bearing Lengths.	N. or S. Distance.	E. or W. Distance.
1	0·83	0·56	51	42·28	28·52	1	0·82	0·57	51	41·78	29·25
2	1·66	1·12	52	43·11	29·08	2	1·64	1·15	52	42·60	29·83
3	2·49	1·68	53	43·94	29·64	3	2·46	1·72	53	43·41	30·40
4	3·32	2·24	54	44·77	30·20	4	3·28	2·29	54	44·23	30·97
5	4·15	2·80	55	45·60	30·76	5	4·10	2·87	55	45·05	31·55
6	4·97	3·36	56	46·43	31·31	6	4·91	3·44	56	45·87	32·12
7	5·80	3·91	57	47·26	31·87	7	5·73	4·02	57	46·69	32·69
8	6·63	4·47	58	48·08	32·43	8	6·55	4·59	58	47·51	33·27
9	7·46	5·03	59	48·91	32·99	9	7·37	5·16	59	48·33	33·84
10	8·29	5·59	60	49·74	33·55	10	8·19	5·74	60	49·15	34·41
11	9·12	6·15	61	50·57	34·11	11	9·01	6·31	61	49·97	34·99
12	9·95	6·71	62	51·40	34·67	12	9·83	6·88	62	50·79	35·56
13	10·78	7·27	63	52·23	35·23	13	10·65	7·46	63	51·61	36·14
14	11·61	7·83	64	53·06	35·79	14	11·47	8·03	64	52·43	36·71
15	12·44	8·39	65	53·89	36·35	15	12·29	8·60	65	53·24	37·28
16	13·26	8·95	66	54·72	36·91	16	13·11	9·18	66	54·06	37·85
17	14·09	9·51	67	55·55	37·47	17	13·93	9·75	67	54·88	38·43
18	14·92	10·07	68	56·37	38·02	18	14·74	10·32	68	55·70	39·00
19	15·75	10·62	69	57·20	38·58	19	15·56	10·90	69	56·52	39·58
20	16·58	11·18	70	58·03	39·14	20	16·38	11·47	70	57·34	40·15
21	17·41	11·74	71	58·86	39·70	21	17·20	12·05	71	58·16	40·72
22	18·24	12·30	72	59·69	40·26	22	18·02	12·62	72	58·98	41·30
23	19·07	12·86	73	60·52	40·82	23	18·84	13·19	73	59·80	41·87
24	19·90	13·42	74	61·35	41·38	24	19·66	13·77	74	60·62	42·44
25	20·73	13·98	75	62·18	41·94	25	20·48	14·34	75	61·44	43·02
26	21·56	14·54	76	63·01	42·50	26	21·30	14·91	76	62·26	43·59
27	22·38	15·10	77	63·84	43·06	27	22·12	15·49	77	63·07	44·17
28	23·21	15·66	78	64·67	43·62	28	22·94	16·06	78	63·89	44·74
29	24·04	16·22	79	65·49	44·18	29	23·76	16·63	79	64·71	45·31
30	24·87	16·78	80	66·32	44·74	30	24·57	17·21	80	65·53	45·89
31	25·70	17·33	81	67·15	45·29	31	25·39	17·78	81	66·35	46·46
32	26·53	17·89	82	67·98	45·85	32	26·21	18·35	82	67·17	47·03
33	27·36	18·45	83	68·81	46·41	33	27·03	18·93	83	67·99	47·61
34	28·19	19·01	84	69·64	46·97	34	27·85	19·50	84	68·81	48·18
35	29·02	19·57	85	70·47	47·53	35	28·67	20·08	85	69·63	48·75
36	29·85	20·13	86	71·30	48·09	36	29·49	20·65	86	70·45	49·33
37	30·67	20·69	87	72·13	48·65	37	30·31	21·22	87	71·27	49·90
38	31·50	21·25	88	72·96	49·21	38	31·13	21·80	88	72·09	50·48
39	32·33	21·81	89	73·78	49·77	39	31·95	22·37	89	72·90	51·05
40	33·16	22·37	90	74·61	50·33	40	32·77	22·94	90	73·72	51·62
41	33·99	22·93	91	75·44	50·89	41	33·59	23·52	91	74·54	52·20
42	34·82	23·49	92	76·27	51·45	42	34·40	24·09	92	75·36	52·77
43	35·65	24·05	93	77·10	52·00	43	35·22	24·66	93	76·18	53·34
44	36·48	24·60	94	77·93	52·56	44	36·04	25·24	94	77·00	53·92
45	37·31	25·16	95	78·76	53·12	45	36·86	25·81	95	77·82	54·49
46	38·14	25·72	96	79·59	53·68	46	37·68	26·38	96	78·64	55·06
47	38·96	26·28	97	80·42	54·24	47	38·50	26·96	97	79·46	55·64
48	39·79	26·84	98	81·25	54·80	48	39·32	27·53	98	80·28	56·21
49	40·62	27·40	99	82·07	55·36	49	40·14	28·11	99	81·10	56·78
50	41·45	27·96	100	82·90	55·92	50	40·96	28·68	100	81·92	57·36
	E. or W.	N. or S.		E. or W.	N. or S.		E. or W.	N. or S.		E. or W.	N. or S.
56°						55°					

36°						37°					
Bearing Lengths.	N. or S. Distance.	E. or W. Distance.	Bearing Lengths.	N. or S. Distance.	E. or W. Distance.	Bearing Lengths.	N. or S. Distance.	E. or W. Distance.	Bearing Lengths.	N. or S. Distance.	E. or W. Distance.
1	0·81	0·59	51	41·26	29·98	1	0·80	0·60	51	40·73	30·69
2	1·62	1·18	52	42·07	30·57	2	1·60	1·20	52	41·53	31·29
3	2·43	1·76	53	42·88	31·15	3	2·40	1·81	53	42·33	31·90
4	3·24	2·35	54	43·69	31·74	4	3·19	2·41	54	43·13	32·50
5	4·05	2·94	55	44·50	32·33	5	3·99	3·01	55	43·93	33·10
6	4·85	3·53	56	45·31	32·92	6	4·79	3·61	56	44·72	33·70
7	5·66	4·12	57	46·11	33·50	7	5·59	4·21	57	45·52	34·30
8	6·47	4·70	58	46·92	34·09	8	6·39	4·81	58	46·32	34·90
9	7·28	5·29	59	47·73	34·68	9	7·19	5·42	59	47·12	35·51
10	8·09	5·88	60	48·54	35·27	10	7·99	6·02	60	47·92	36·11
11	8·90	6·47	61	49·35	35·86	11	8·79	6·62	61	48·72	36·71
12	9·71	7·05	62	50·16	36·44	12	9·58	7·22	62	49·52	37·31
13	10·52	7·64	63	50·97	37·03	13	10·38	7·82	63	50·31	37·91
14	11·33	8·23	64	51·78	37·62	14	11·18	8·43	64	51·11	38·52
15	12·14	8·82	65	52·59	38·21	15	11·98	9·03	65	51·91	39·12
16	12·94	9·40	66	53·40	38·79	16	12·78	9·63	66	52·71	39·72
17	13·75	9·99	67	54·20	39·38	17	13·58	10·23	67	53·51	40·32
18	14·56	10·58	68	55·01	39·97	18	14·38	10·83	68	54·31	40·92
19	15·37	11·17	69	55·82	40·56	19	15·17	11·43	69	55·11	41·52
20	16·18	11·76	70	56·63	41·14	20	15·97	12·04	70	55·90	42·13
21	16·99	12·34	71	57·44	41·73	21	16·77	12·64	71	56·70	42·73
22	17·80	12·93	72	58·25	42·32	22	17·57	13·24	72	57·50	43·33
23	18·61	13·52	73	59·06	42·91	23	18·37	13·84	73	58·30	43·93
24	19·42	14·11	74	59·87	43·50	24	19·17	14·44	74	59·10	44·53
25	20·23	14·69	75	60·68	44·08	25	19·97	15·05	75	59·90	45·14
26	21·03	15·28	76	61·49	44·67	26	20·76	15·65	76	60·70	45·74
27	21·84	15·87	77	62·29	45·26	27	21·56	16·25	77	61·50	46·34
28	22·65	16·46	78	63·10	45·85	28	22·36	16·85	78	62·29	46·94
29	23·46	17·05	79	63·91	46·44	29	23·16	17·45	79	63·09	47·54
30	24·27	17·63	80	64·72	47·02	30	23·96	18·05	80	63·89	48·14
31	25·08	18·22	81	65·53	47·61	31	24·76	18·66	81	64·69	48·75
32	25·89	18·81	82	66·34	48·20	32	25·56	19·26	82	65·49	49·35
33	26·70	19·40	83	67·15	48·79	33	26·36	19·86	83	66·29	49·95
34	27·51	19·98	84	67·96	49·37	34	27·15	20·46	84	67·08	50·55
35	28·32	20·57	85	68·77	49·96	35	27·95	21·06	85	67·88	51·15
36	29·12	21·16	86	69·58	50·55	36	28·75	21·67	86	68·68	51·76
37	29·93	21·75	87	70·38	51·14	37	29·55	22·27	87	69·48	52·36
38	30·74	22·34	88	71·19	51·73	38	30·35	22·87	88	70·28	52·96
39	31·55	22·92	89	72·00	52·31	39	31·15	23·47	89	71·08	53·56
40	32·36	23·51	90	72·81	52·90	40	31·95	24·07	90	71·88	54·16
41	33·17	24·10	91	73·62	53·49	41	32·74	24·67	91	72·68	54·76
42	33·98	24·69	92	74·43	54·08	42	33·54	25·28	92	73·47	55·37
43	34·79	25·27	93	75·24	54·66	43	34·34	25·88	93	74·27	55·97
44	35·60	25·86	94	76·05	55·25	44	35·14	26·48	94	75·07	56·57
45	36·41	26·45	95	76·86	55·84	45	35·94	27·08	95	75·87	57·17
46	37·21	27·04	96	77·67	56·43	46	36·74	27·68	96	76·67	57·77
47	38·02	27·63	97	78·47	57·02	47	37·54	28·29	97	77·47	58·38
48	38·83	28·21	98	79·28	57·60	48	38·33	28·89	98	78·27	58·98
49	39·64	28·80	99	80·09	58·19	49	39·13	29·49	99	79·07	59·58
50	40·45	29·39	100	80·90	58·78	50	39·93	30·09	100	79·86	60·18
	E. or W.	N. or S.		E. or W.	N. or S.		E. or W.	N. or S.		E. or W.	N. or S.
54°						53°					

38°						39°					
Bearing Lengths.	N. or S. Distance.	E. or W. Distance.	Bearing Lengths.	N. or S. Distance.	E. or W. Distance.	Bearing Lengths.	N. or S. Distance.	E. or W. Distance.	Bearing Lengths.	N. or S. Distance.	E. or W. Distance.
1	0·79	0·62	51	40·19	31·40	1	0·78	0·63	51	39·63	32·09
2	1·58	1·23	52	40·98	32·01	2	1·55	1·26	52	40·41	32·72
3	2·36	1·85	53	41·76	32·63	3	2·33	1·89	53	41·19	33·35
4	3·15	2·46	54	42·55	33·25	4	3·11	2·52	54	41·97	33·98
5	3·94	3·08	55	43·34	33·86	5	3·89	3·15	55	42·74	34·61
6	4·73	3·69	56	44·13	34·48	6	4·66	3·78	56	43·52	35·24
7	5·52	4·31	57	44·92	35·09	7	5·44	4·41	57	44·30	35·87
8	6·30	4·93	58	45·70	35·71	8	6·22	5·03	58	45·07	36·50
9	7·09	5·54	59	46·49	36·32	9	6·99	5·66	59	45·85	37·13
10	7·88	6·16	60	47·28	36·94	10	7·77	6·29	60	46·63	37·76
11	8·67	6·77	61	48·07	37·56	11	8·55	6·92	61	47·41	38·39
12	9·46	7·39	62	48·86	38·17	12	9·33	7·55	62	48·18	39·02
13	10·24	8·00	63	49·64	38·79	13	10·10	8·18	63	48·96	39·65
14	11·03	8·62	64	50·43	39·40	14	10·88	8·81	64	49·74	40·28
15	11·82	9·23	65	51·22	40·02	15	11·66	9·44	65	50·51	40·91
16	12·61	9·85	66	52·01	40·63	16	12·43	10·07	66	51·29	41·54
17	13·40	10·47	67	52·80	41·25	17	13·21	10·70	67	52·07	42·16
18	14·18	11·08	68	53·58	41·86	18	13·99	11·33	68	52·85	42·79
19	14·97	11·70	69	54·37	42·48	19	14·77	11·96	69	53·62	43·42
20	15·76	12·31	70	55·16	43·10	20	15·54	12·59	70	54·40	44·05
21	16·55	12·93	71	55·95	43·71	21	16·32	13·22	71	55·18	44·68
22	17·34	13·54	72	56·74	44·33	22	17·10	13·84	72	55·95	45·31
23	18·12	14·16	73	57·52	44·94	23	17·87	14·47	73	56·73	45·94
24	18·91	14·78	74	58·31	45·56	24	18·65	15·10	74	57·51	46·57
25	19·70	15·39	75	59·10	46·17	25	19·43	15·73	75	58·29	47·20
26	20·49	16·01	76	59·89	46·79	26	20·21	16·36	76	59·06	47·83
27	21·28	16·62	77	60·68	47·41	27	20·98	16·99	77	59·84	48·46
28	22·06	17·24	78	61·46	48·02	28	21·76	17·62	78	60·62	49·09
29	22·85	17·85	79	62·25	48·64	29	22·54	18·25	79	61·39	49·72
30	23·64	18·47	80	63·04	49·25	30	23·31	18·88	80	62·17	50·34
31	24·43	19·09	81	63·83	49·87	31	24·09	19·51	81	62·95	50·97
32	25·22	19·70	82	64·62	50·48	32	24·87	20·14	82	63·73	51·60
33	26·00	20·32	83	65·40	51·10	33	25·65	20·77	83	64·50	52·23
34	26·79	20·93	84	66·19	51·72	34	26·42	21·40	84	65·28	52·86
35	27·58	21·55	85	66·98	52·33	35	27·20	22·03	85	66·06	53·49
36	28·37	22·16	86	67·77	52·95	36	27·98	22·66	86	66·83	54·12
37	29·16	22·78	87	68·56	53·56	37	28·75	23·28	87	67·61	54·75
38	29·94	23·40	88	69·34	54·18	38	29·53	23·91	88	68·39	55·38
39	30·73	24·01	89	70·13	54·79	39	30·31	24·54	89	69·17	56·01
40	31·52	24·63	90	70·92	55·41	40	31·09	25·17	90	69·94	56·64
41	32·31	25·24	91	71·71	56·03	41	31·86	25·80	91	70·72	57·27
42	33·10	25·86	92	72·50	56·64	42	32·64	26·43	92	71·50	57·90
43	33·88	26·47	93	73·28	57·26	43	33·42	27·06	93	72·27	58·53
44	34·67	27·09	94	74·07	57·87	44	34·19	27·69	94	73·05	59·16
45	35·46	27·70	95	74·86	58·49	45	34·97	28·32	95	73·83	59·78
46	36·25	28·32	96	75·65	59·10	46	35·75	28·95	96	74·61	60·41
47	37·04	28·94	97	76·44	59·72	47	36·53	29·58	97	75·38	61·04
48	37·82	29·55	98	77·22	60·33	48	37·30	30·21	98	76·16	61·67
49	38·61	30·17	99	78·01	60·95	49	38·08	30·84	99	76·94	62·30
50	39·40	30·78	100	78·80	61·57	50	38·86	31·47	100	77·72	62·93
	E. or W.	N. or S.		E. or W.	N. or S.		E. or W.	N. or S.		E. or W.	N. or S.
52°						51°					

40°					
Bearing Lengths.	N. or S. Distance.	E. or W. Distance.	Bearing Lengths.	N. or S. Distance.	E. or W. Distance.
1	0·77	0·64	51	39·07	32·78
2	1·33	1·29	52	39·83	33·43
3	2·30	1·93	53	40·60	34·07
4	3·06	2·57	54	41·37	34·71
5	3·81	3·21	55	42·13	35·35
6	4·60	3·86	56	42·90	36·00
7	5·36	4·50	57	43·66	36·64
8	6·13	5·14	58	44·43	37·28
9	6·89	5·79	59	45·20	37·93
10	7·66	6·43	60	45·96	38·57
11	8·43	7·07	61	46·73	39·21
12	9·19	7·71	62	47·49	39·85
13	9·96	8·36	63	48·26	40·50
14	10·72	9·00	64	49·03	41·14
15	11·49	9·64	65	49·79	41·78
16	12·26	10·28	66	50·56	42·43
17	13·02	10·93	67	51·32	43·07
18	13·79	11·57	68	52·09	43·71
19	14·55	12·21	69	52·86	44·35
20	15·32	12·86	70	53·62	45·00
21	16·09	13·50	71	54·39	45·64
22	16·85	14·14	72	55·15	46·28
23	17·62	14·78	73	55·92	46·92
24	18·38	15·43	74	56·69	47·57
25	19·15	16·07	75	57·45	48·21
26	19·92	16·71	76	58·22	48·85
27	20·68	17·36	77	58·99	49·49
28	21·45	18·00	78	59·75	50·14
29	22·21	18·64	79	60·52	50·78
30	22·98	19·28	80	61·30	51·42
31	23·75	19·03	81	62·05	52·07
32	24·51	20·57	82	62·82	52·71
33	25·28	21·21	83	63·58	53·35
34	26·05	21·85	84	64·35	53·99
35	26·81	22·50	85	65·11	54·64
36	27·58	23·14	86	65·88	55·28
37	28·34	23·78	87	66·65	55·92
38	29·11	24·43	88	67·41	56·57
39	29·88	25·07	89	68·18	57·21
40	30·64	25·71	90	68·94	57·85
41	31·41	26·35	91	69·71	58·49
42	32·17	27·00	92	70·48	59·14
43	32·94	27·64	93	71·24	59·78
44	33·71	28·28	94	72·01	60·42
45	34·47	28·93	95	72·77	61·07
46	35·24	29·57	96	73·54	61·71
47	36·00	30·21	97	74·31	62·35
48	36·77	30·85	98	75·07	62·99
49	37·54	31·50	99	75·84	63·64
50	38·18	32·14	100	76·60	64·28
	E. or W.	N. or S.		E. or W.	N. or S.
50°					

41°					
Bearing Lengths.	N. or S. Distance.	E. or W. Distance.	Bearing Lengths.	N. or S. Distance.	E. or W. Distance.
1	0·75	0·66	51	38·49	33·46
2	1·51	1·31	52	39·24	34·12
3	2·26	1·97	53	40·00	34·77
4	3·02	2·62	54	40·75	35·43
5	3·77	3·28	55	41·51	36·08
6	4·53	3·94	56	42·26	36·74
7	5·28	4·59	57	43·02	37·40
8	6·04	5·25	58	13·77	38·05
9	6·79	5·00	59	44·53	38·71
10	7·55	6·56	60	45·28	39·36
11	8·30	7·22	61	46·04	40·02
12	9·06	7·87	62	46·79	40·68
13	9·81	8·53	63	47·55	41·33
14	10·57	9·18	64	48·30	41·99
15	11·32	9·84	65	49·06	42·64
16	12·08	10·50	66	49·81	43·30
17	12·83	11·15	67	50·57	43·96
18	13·58	11·81	68	51·32	44·61
19	14·34	12·47	69	52·07	45·27
20	15·09	13·12	70	52·83	45·92
21	15·85	13·78	71	53·58	46·58
22	16·00	14·43	72	54·34	47·24
23	17·36	15·09	73	55·09	47·89
24	18·11	15·75	74	55·85	48·55
25	18·87	16·40	75	56·60	49·20
26	19·02	17·06	76	57·36	49·86
27	20·38	17·71	77	58·11	50·52
28	21·13	18·37	78	58·87	51·17
29	21·89	19·03	79	59·62	51·83
30	22·64	19·68	80	60·38	52·48
31	23·40	20·34	81	61·13	53·14
32	24·15	20·99	82	61·89	53·80
33	24·91	21·65	83	62·64	54·45
34	25·66	22·31	84	63·40	55·11
35	26·41	22·96	85	64·15	55·77
36	27·17	23·62	86	64·91	56·42
37	27·92	24·27	87	65·66	57·08
38	28·68	24·93	88	66·41	57·73
39	29·43	25·59	89	67·17	58·39
40	30·19	26·24	90	67·92	59·05
41	30·94	26·90	91	68·68	59·70
42	31·70	27·55	92	69·43	60·36
43	32·45	28·21	93	70·19	61·01
44	33·21	28·87	94	70·94	61·67
45	33·96	29·52	95	71·70	62·33
46	34·72	30·18	96	72·45	62·98
47	35·47	30·83	97	73·21	63·64
48	36·23	31·49	98	73·96	64·29
49	36·98	32·15	99	74·72	64·95
50	37·74	32·80	100	75·47	65·61
	E. or W.	N. or S.		E. or W.	N. or S.
49°					

42°						43°					
Bearing Lengths.	N. or S. Distance.	E. or W. Distance.	Bearing Lengths.	N. or S. Distance.	E. or W. Distance.	Bearing Lengths.	N. or S. Distance.	E. or W. Distance.	Bearing Lengths.	N. or S. Distance.	E. or W. Distance.
1	0·74	0·67	51	37·90	34·13	1	0·73	0·68	51	37·30	34·78
2	1·49	1·34	52	38·64	34·79	2	1·46	1·36	52	38·03	35·46
3	2·23	2·01	53	39·39	35·46	3	2·19	2·05	53	38·76	36·15
4	2·97	2·68	54	40·13	36·13	4	2·93	2·73	54	39·49	36·83
5	3·72	3·36	55	40·87	36·80	5	3·66	3·41	55	40·22	37·51
6	4·46	4·01	56	41·62	37·47	6	4·39	4·09	56	40·96	38·19
7	5·20	5·68	57	42·36	38·14	7	5·12	4·77	57	41·69	38·87
8	5·95	5·35	58	43·10	38·81	8	5·85	5·46	58	42·42	39·56
9	6·69	6·02	59	43·85	39·48	9	6·58	6·14	59	43·15	40·24
10	7·43	6·69	60	44·59	40·15	10	7·31	0·82	60	43·88	40·92
11	8·17	7·36	61	45·33	40·82	11	8·04	7·50	61	44·61	41·60
12	9·92	8·03	62	46·07	41·49	12	8·78	8·18	62	45·34	42·28
13	9·66	8·70	63	46·82	42·16	13	9·51	8·87	63	46·08	42·97
14	10·40	9·37	64	47·56	42·82	14	10·24	9·55	64	46·81	43·65
15	11·16	10·04	65	48·30	43·49	15	10·97	10·23	65	47·54	44·33
16	11·89	10·71	66	49·05	44·16	16	11·70	10·91	66	48·27	45·01
17	12·63	11·38	67	49·79	44·83	17	12·43	11·59	67	49·00	45·69
18	13·38	12·04	68	50·53	45·50	18	13·16	12·28	68	49·73	46·38
19	14·12	12·71	69	51·28	46·17	19	13·90	12·96	69	50·40	47·06
20	14·86	13·38	70	52·02	46·84	20	14·63	13·64	70	51·19	47·74
21	15·61	14·05	71	52·76	47·51	21	15·36	14·32	71	51·93	48·42
22	16·35	14·72	72	53·51	48·18	22	16·09	15·00	72	52·66	49·10
23	17·09	15·39	73	54·25	48·85	23	16·82	15·69	73	53·30	49·79
24	17·84	16·06	74	54·99	49·52	24	17·55	16·37	74	54·12	50·47
25	18·58	16·73	75	55·69	49·79	25	18·28	17·05	75	54·85	51·15
26	19·32	17·40	76	56·48	50·85	26	19·02	17·73	76	55·58	51·83
27	20·06	18·07	77	57·22	51·52	27	19·75	18·41	77	56·31	52·51
28	20·81	18·74	78	57·96	52·19	28	20·48	19·10	78	57·05	53·20
29	21·55	19·40	79	58·71	52·86	29	21·21	19·78	79	57·78	53·88
30	22·29	20·07	80	59·45	53·53	30	21·94	20·46	80	58·51	54·56
31	23·04	20·74	81	60·19	54·20	31	22·67	21·14	81	59·24	55·24
32	23·78	21·41	82	60·94	54·87	32	23·40	21·82	82	59·97	55·92
33	24·52	22·08	83	61·68	55·54	33	24·13	22·51	83	60·70	56·61
34	25·27	22·75	84	62·42	56·21	34	24·87	23·19	84	61·43	57·29
35	26·01	23·42	85	63·17	56·88	35	25·60	23·87	85	62·16	57·97
36	26·75	24·09	86	63·91	57·55	36	26·33	24·55	86	62·90	58·65
37	27·50	24·76	87	64·65	58·21	37	27·06	25·23	87	63·63	59·33
38	28·24	25·43	88	65·40	58·88	38	27·79	25·92	88	64·36	60·02
39	28·98	26·10	89	66·14	59·55	39	28·52	26·60	89	65·09	60·70
40	29·73	26·80	90	66·88	60·22	40	29·25	27·28	90	65·82	61·38
41	30·47	27·43	91	67·63	60·89	41	29·99	27·96	91	66·55	61·06
42	31·21	28·10	92	68·37	61·56	42	30·72	28·64	92	67·28	62·74
43	31·96	28·77	93	69·11	62·23	43	31·45	29·33	93	68·02	63·43
44	32·70	29·44	94	69·85	62·90	44	32·18	30·01	94	68·75	64·11
45	33·44	30·11	95	70·60	63·57	45	32·91	30·69	95	69·48	64·79
46	34·18	30·78	96	71·34	64·24	46	33·64	31·37	96	70·21	65·47
47	34·93	31·45	97	72·08	64·91	47	34·37	32·05	97	70·94	66·15
48	35·67	32·12	98	72·83	65·57	48	35·10	32·74	98	71·67	66·84
49	36·41	32·79	99	73·57	66·24	49	35·84	33·42	99	72·40	67·52
50	37·20	33·50	100	74·31	66·91	50	36·57	34·10	100	73·14	68·20
	E. or W.	N. or S.		E. or W.	N. or S.		E. or W.	N. or S.		E. or W.	N. or S.
48°						47°					

44°					
Bearing Lengths.	N. or S. Distance.	E. or W. Distance.	Bearing Lengths.	N. or S. Distance.	E. or W. Distance.
1	0·72	0·69	51	36·69	35·43
2	1·44	1·39	52	37·41	36·12
3	2·16	2·08	53	38·13	36·82
4	2·88	2·78	54	38·84	37·51
5	3·60	3·47	55	39·56	38·21
6	4·32	4·17	56	40·28	38·90
7	5·04	4·86	57	41·00	39·60
8	5·75	5·55	58	41·72	40·29
9	6·47	6·25	59	42·44	40·98
10	7·19	6·95	60	43·16	41·68
11	7·91	7·64	61	43·88	42·37
12	8·63	8·34	62	44·60	42·07
13	9·35	9·03	63	45·32	43·76
14	10·07	9·73	64	46·04	44·46
15	10·79	10·42	65	46·76	45·15
16	11·51	11·11	66	47·48	45·85
17	12·23	11·81	67	48·20	46·54
18	12·95	12·50	68	48·92	47·24
19	13·67	13·20	69	49·63	47·93
20	14·39	13·89	70	50·35	48·63
21	15·11	14·59	71	51·07	49·32
22	15·83	15·28	72	51·79	50·02
23	16·54	15·98	73	52·51	50·71
24	17·26	16·67	74	53·23	51·40
25	17·98	17·37	75	53·95	52·10
26	18·70	18·06	76	54·67	52·79
27	19·42	18·76	77	55·39	53·49
28	20·14	19·45	78	56·11	54·18
29	20·86	20·15	79	56·83	54·88
30	21·58	20·84	80	57·55	55·57
31	22·30	21·53	81	58·27	56·27
32	23·02	22·23	82	58·99	56·96
33	23·74	22·92	83	59·71	57·66
34	24·46	23·62	84	60·42	58·35
35	25·18	24·31	85	61·14	59·05
36	25·90	25·01	86	61·86	59·74
37	26·62	25·70	87	62·58	60·44
38	27·33	26·40	88	63·30	61·13
39	28·05	27·09	89	64·02	61·82
40	28·77	27·79	90	64·74	62·52
41	29·49	28·48	91	65·46	63·21
42	30·21	29·18	92	66·18	63·91
43	30·93	29·87	93	66·90	64·60
44	31·65	30·57	94	67·62	65·30
45	32·37	31·26	95	68·34	65·99
46	33·09	31·95	96	69·06	66·69
47	33·81	32·65	97	69·78	67·38
48	34·53	33·34	98	70·50	68·08
49	35·25	34·04	99	71·21	68·77
50	35·97	34·73	100	71·93	69·47
	E. or W.	N. or S.		E. or W.	N. or S.
46°					

45°					
Bearing Lengths.	N. or S. Distance.	E. or W. Distance.	Bearing Lengths.	N. or S. Distance.	E. or W. Distance.
1	0·71	0·71	51	36·06	36·06
2	1·41	1·41	52	36·77	36·77
3	2·12	2·12	53	37·48	37·48
4	2·83	2·83	54	38·18	38·18
5	3·54	3·54	55	38·89	38·89
6	4·24	4·24	56	39·60	39·60
7	4·95	4·95	57	40·31	40·31
8	5·66	5·66	58	41·01	41·01
9	6·36	6·36	59	41·72	41·72
10	7·07	7·07	60	42·43	42·43
11	7·78	7·78	61	43·13	43·13
12	8·49	8·49	62	43·84	43·84
13	9·19	9·19	63	44·55	44·55
14	9·90	9·90	64	45·26	45·26
15	10·61	10·61	65	45·96	45·96
16	11·31	11·31	66	46·67	46·67
17	12·02	12·02	67	47·38	47·38
18	12·73	12·73	68	48·08	48·08
19	13·44	13·44	69	48·79	48·79
20	14·14	14·14	70	49·50	49·50
21	14·85	14·85	71	50·20	50·20
22	15·56	15·56	72	50·91	50·91
23	16·26	16·26	73	51·62	51·62
24	16·97	16·97	74	52·33	52·33
25	17·68	17·68	75	53·03	53·03
26	18·38	18·38	76	53·74	53·74
27	19·09	19·09	77	54·45	54·45
28	19·80	19·80	78	55·15	55·15
29	20·51	20·51	79	55·86	55·86
30	21·21	21·21	80	56·57	56·57
31	21·92	21·92	81	57·28	57·28
32	22·63	22·63	82	57·98	57·98
33	23·33	23·33	83	58·69	58·69
34	24·04	24·04	84	59·40	59·40
35	24·75	24·75	85	60·10	60·10
36	25·46	25·46	86	60·81	60·81
37	26·16	26·16	87	61·52	61·52
38	26·87	26·87	88	62·23	62·23
39	27·58	27·58	89	62·93	62·93
40	28·28	28·28	90	63·64	63·64
41	28·99	28·99	91	64·33	64·33
42	29·70	29·70	92	65·05	65·05
43	30·41	30·41	93	65·76	65·76
44	31·11	31·11	94	66·47	66·47
45	31·82	31·82	95	67·18	67·18
46	32·53	32·53	96	67·88	67·88
47	33·23	33·23	97	68·59	68·59
48	33·94	33·94	98	69·30	69·30
49	34·65	34·65	99	70·00	70·00
50	35·35	35·35	100	70·71	70·71
	E. or W.	N. or S.		E. or W.	N. or S.
45°					

OF THE PRODUCE OF SEAMS OF COAL.

(61.) From the various experiments which have been made on the produce of tracts of coal mines, in the neighbourhood of Newcastle-upon-Tyne, it has been found that a cubic yard of coal weighs ·936 of a ton; therefore, an acre of that stratum, 1 foot thick, will produce (if all wrought out) 1510 tons; consequently an equal area of stratum, 2, 3, 4, &c., feet in thickness, will produce 2, 3, 4, &c., times the quantity of tons of coal that a seam of 1 foot thick will produce.

From this datum easy rules may be constructed for the use of the practical miner, which with facility may be retained for application in calculating the produce of seams of any given thickness in tons.

To find the number of tons of coal contained per acre by a seam of any given thickness.

Rule I.—Multiply 1510 by the thickness or height of the seam in feet, and the product will be the number of tons of coal contained in an acre of that seam.

To find the number of tons of coal produced per acre by a seam, where part thereof is only worked or taken away, the other part being left as a support to the roof.

Rule II.—As the sum of the two parts, *i.e.*, that left and that taken away, is to the part excavated or taken away, so is the whole number of tons contained in an acre of the seam to the number of tons produced per acre by the excavated part.

Example I.—What number of tons of coal is contained in an acre of coal stratum 6 feet thick?

From rule 1st, $1510 \times 6 = 9060$ tons, the content.

Example II.—What number of tons of coal is contained in an area of coal stratum of 100 acres, 5 feet thick?

$1510 \times 5 = 7550$ tons contained in one acre.

Then $7550 \times 100 = 755{,}000$ tons contained in 100 acres.

Example III.—What number of tons of coal is contained in 400 acres of coal stratum 5 feet 3 inches thick?

First 5 ft. 3 in. $= \frac{21}{4}$ ft.

And $\frac{1510 \times 21 \times 400}{4} = 3{,}171{,}000$ tons.

Example IV.—What number of tons of coal is contained in 400 acres of coal stratum 8 feet 4 inches thick?

First 8 ft. 4 in. $= \frac{25}{3}$ ft.

And $\frac{1510 \times 25 \times 400}{3} = 5{,}033{,}333$ tons.

Example V.—What number of tons of coal is contained in 500 acres of coal stratum 4 feet 9 inches thick, excluding a band of stone which lies therein 6 inches thick?

$4{\cdot}75 - {\cdot}50 = 4{\cdot}25$ feet, the thickness of the coal stratum, exclusive of the band of stone.

Then $1510 \times 4{\cdot}25 \times 500 = 3{,}208{,}750$ tons contained in 500 acres.

Example VI.—In a seam of coal which is 7 feet 3 inches thick; that is to say, 6 feet of its thickness is marketable, and 1 foot 3 inches inferior; I wish to know the produce in tons per acre, both of the marketable and the inferior parts of the seam?

$1510 \times 6 = 9060$ tons per acre, the marketable produce of the seam

First 1 ft. 3 in. $= \frac{5}{4}$ ft.

And $\frac{1510 \times 5}{4} = 1887\frac{1}{2}$ tons of the inferior parts.

Example VII.—In a seam of coal 6 feet thick, I wish to know what number of tons it produces per acre, when 1 part is taken away, and 2 left for pillars or supports?

$1510 \times 6 = 9060$ tons, the whole content per acre.

From rule 2nd, as $2 + 1 = 3 : 1 :: 9060 : 3020$ tons, the produce per acre of the part taken away.

EXAMPLE VIII.—In a seam of coal 3 feet 6 inches thick, I wish to know what number of tons it will produce per acre, when two parts are taken away and 1 left?

$1510 \times 3{\cdot}5 = 5285$ tons, the content per acre.

As $2 + 1 = 3 : 2 :: 5285 : 3523{\cdot}33$ tons, the produce per acre of the part taken away.

EXAMPLE IX.—In 1000 acres of coal 5 feet thick, whereof 2 parts are worked and 1 left, I wish to know how many years this stratum of coal will produce an annual quantity of 50,000 tons?

$1510 \times 5 = 7550$ tons, the whole produce of the seam per acre.

Then as $3 : 2 :: 7550 : 5033$, the quantity got per acre.

And $\dfrac{5033 \times 1000}{50{,}000} = 100{\cdot}66$ years.

EXAMPLE X.—I have a tract of 600 acres of coal stratum, containing 2 seams, the first 5 feet 3 inches thick, and the second 3 feet 6 inches thick: Out of the first seam 3 parts are got and 1 left; and out of the second 4 parts are got and 1 left. Now, if the annual vend of the two seams together is 75,000 tons, what number must be wrought out of each seam yearly, so that they may terminate together; and how many years will the colliery last?

$\dfrac{1510 \times 21}{4} = 7927\frac{1}{2}$ tons, the whole produce per acre of the first seam.

And $4 : 3 :: 7927\frac{1}{2} : 5945$ tons, the quantity wrought per acre out of the first seam.

Then $5945 \times 600 = 3{,}567{,}000$ tons, total produce of the first seam.

Again, $\dfrac{1510 \times 7}{2} = 5285$ tons, the whole produce per acre of the second seam.

And 5 : 4 : : 5285 : 4228 tons, the quantity wrought per acre out of the second seam.

Then 4228 × 600 = 2,536,800 tons, total produce of the second seam.

Now, to make the two seams terminate together, the quantity wrought out of each seam annually must bear the same proportion to each other as the quantity wrought out of each acre of each seam.

Therefore 5945 + 4228 = 10,173 : 5945 : : 75,000 : 43,829 tons, the quantity to be wrought out of the first seam annually.

And 75000 − 43829 = 31,171 tons, the quantity to be wrought out of the second seam annually.

Whence $\frac{3567000 + 2536800}{75000}$ = 80 years 18 days, the duration of the colliery.

Note.—Elaborate statistics and details of the extent, the probable produce and duration of all the coal-fields in the United Kingdom, also the extent and thickness of the strata of those of the United States and the British colonies, as far as they are known; as well as those of Belgium, France, Germany, and other foreign countries, are given in the *Reports of the Institution of Mining Engineers of Newcastle-upon-Tyne*, to which the student is referred, who may be desirous to be acquainted with these subjects.

Questions in Mine Surveying.

Note.—The solutions to the two following Questions in Mine Surveying will require a knowledge of the application of Algebra to Geometry, and the latter of the two will require a further knowledge of the application of Spherical Trigonometry to Astronomy. See *Question XVII, page* 211, *Baker's Land and Engineering Surveying. Weale's Series.* The student will have no difficulty in sketching the figures and assigning the dimensions to the given parts in the two Questions.

Question 1.—There are four drifts in a coal mine forming a trapezium, the given lengths of which are a, b, c, d, and the sums of the opposite angles of the trapezium are known to be equal to two right angles, none of the angles being separately given. It is required to plot this subterraneous survey by the help of the following formula.

Let S = half the sum of the lengths of the four drifts = $\frac{1}{2}(a + b + c + d)$, and D the diameter of circle, which will circumscribe the trapezium; then

$$D = \sqrt{\left\{\frac{(ac + bd)(ab + cd)(ad + bc)}{(S-a)(S-b)(S-c)(S-d)}\right\}}$$

Note.—This formula will divest the preceding Question of its chief difficulty, while it will accustom the student to the application of this species of mathematical analysis.

Question 2.—There are five straight drifts, AB, BC, CD, DE, EA, in a coal-mine, forming an irregular polygon; now, the several lengths of each of the five drifts are given, and the angles at B, C, and D are known to be equal to one another, but are not given: also at each of the angles B and D is a shaft, and the tops of these two shafts range with the sun at 3 h. 35′ P.M. on the 22d of October, 1860. It is required from these data to plot this subterraneous survey in its true position with respect to the cardinal points.

NOTE.—This question was proposed by B. Gompertz, Esq., F.R.S., in the *Gentleman's Mathematical Companion;* to which he gave a solution in a concise, novel, and ingenious manner by his *Principles of Imaginary Quantities:* other solutions by the ordinary methods were also given to the same problem.

THE END.

MR. WEALE'S PUBLICATIONS FOR 1861.

RUDIMENTARY SERIES.

In demy 12mo, cloth, price 1s.

RUDIMENTARY.—1.—CHEMISTRY, by Professor FOWNES, F.R.S., including Agricultural Chemistry, for the Use of Farmers.

In demy 12mo, with Woodcuts, cloth, price 1s.

RUDIMENTARY.—2.—NATURAL PHILOSOPHY, by CHARLES TOMLINSON.

In demy 12mo, with Woodcuts, cloth, price 1s. 6d.

RUDIMENTARY.—3.—GEOLOGY, by Major-Gen. PORTLOCK, F.R.S., &c.

In demy 12mo, with Woodcuts, cloth, price 2s.

RUDIMENTARY.—4, 5.—MINERALOGY, with Mr. DANA'S Additions. 2 vols. in 1.

In demy 12mo, with Woodcuts, cloth, price 1s.

RUDIMENTARY.—6.—MECHANICS, by CHARLES TOMLINSON.

In demy 12mo, with Woodcuts, cloth, price 1s. 6d.

RUDIMENTARY.—7.—ELECTRICITY, by Sir WILLIAM SNOW HARRIS, F.R.S.

In demy 12mo, with Woodcuts, cloth, price 1s. 6d.

RUDIMENTARY.—7*.—ON GALVANISM; ANIMAL AND VOLTAIC ELECTRICITY; by Sir W. SNOW HARRIS.

In demy 12mo, with Woodcuts, cloth, price 3s. 6d.

RUDIMENTARY.—8, 9, 10—MAGNETISM, Concise Exposition of, by Sir W. SNOW HARRIS, 3 vols. in 1.

In demy 12mo, with Woodcuts, cloth, price 2s.

RUDIMENTARY.—11, 11*.—ELECTRIC TELEGRAPH, History of the, by E. HIGHTON, C.E.

In demy 12mo, with Woodcuts, cloth, price 1s.

RUDIMENTARY.—12.—PNEUMATICS, by CHARLES TOMLINSON.

In demy 12mo, with Woodcuts, cloth, price 4s. 6d.

RUDIMENTARY.—13, 14, 15, 15*.—CIVIL ENGINEERING, by HENRY LAW, C.E., 3 vols.; and Supplement by G. R. BURNELL, C.E.

In demy 12mo, with Woodcuts, cloth, price 1s.

RUDIMENTARY.—16.—ARCHITECTURE, Orders of, by W. H. LEEDS.

In demy 12mo, with Woodcuts, cloth, price 1s. 6d.

RUDIMENTARY.—17.—ARCHITECTURE Styles of, by T. BURY, Architect.

John Weale, 59, High Holborn, London, W.C.

MR. WEALE'S RUDIMENTARY SERIES.

In demy 12mo, cloth, price 1s.

RUDIMENTARY. — 41. — PAINTING ON GLASS, Essay on, by E. O. FROMBERG.

In demy 12mo, with Woodcuts, cloth, price 1s.

RUDIMENTARY.—42.— COTTAGE BUILDING, Treatise on.

In demy 12mo, with Woodcuts, cloth, price 1s.

RUDIMENTARY. — 43. — TUBULAR AND GIRDER BRIDGES, and others, Treatise on, more particularly describing the Britannia and Conway Bridges.

In demy 12mo, with Woodcuts, cloth, price 1s.

RUDIMENTARY.—44.—FOUNDATIONS, &c., by E. DOBSON, C.E.

In demy 12mo, with Woodcuts, cloth, price 1s.

RUDIMENTARY. — 45. — LIMES, CEMENTS, MORTARS, CONCRETE, MASTICS, &c., by G. R. BURNELL, C.E.

In demy 12mo, with Woodcuts, cloth, price 1s.

RUDIMENTARY. — 46. — CONSTRUCTING AND REPAIRING COMMON ROADS, by H. LAW, C.E.

In demy 12mo, with Woodcuts, cloth, price 3s.

RUDIMENTARY. — 47, 48, 49. — CONSTRUCTION AND ILLUMINATION OF LIGHTHOUSES, by ALAN STEVENSON, C.E., 3 vols. in 1.

In demy 12mo, with Woodcuts, cloth, price 1s.

RUDIMENTARY.—50.—LAW OF CONTRACTS FOR WORKS AND SERVICES, by DAVID GIBBONS, S.P.

In demy 12mo, with Woodcuts, cloth, price 3s.

RUDIMENTARY.—51, 52, 53.—NAVAL ARCHITECTURE, Principles of the Science, by J. PEAKE, N.A., 3 vols. in 1.

In demy 12mo, with Woodcuts, cloth, price 1s.

RUDIMENTARY AND ELEMENTARY.—53*. —PRACTICAL CONSTRUCTION concisely stated of Ships for Ocean or River Service, by Captain H. A. SOMMERFELDT, N.R.N.

In royal 4to with Engraved Plates, cloth, price 7s. 6d.

RUDIMENTARY.—53**—ATLAS of 15 Plates to ditto, drawn and engraved to a Scale for Practice.—For the convenience of the Operative Ship Builder the Atlas may be had in three separate Parts. Part I., 2s 6d. Part II., 2s. 6d. Part III., 2s. 6d.

In demy 12mo, with Woodcuts, cloth, price 1s. 6d.

RUDIMENTARY. — 54. — MASTING, MAST-MAKING, AND RIGGING OF SHIPS, by R. KIPPING, N.A.

In demy 12mo, with Woodcuts, cloth, price 2s. 6d.

RUDIMENTARY.—54*.—IRON SHIP BUILDING, by JOHN GRANTHAM, N.A. and C.E.

In demy 12mo, with Woodcuts, cloth, price 2s.

RUDIMENTARY. — 55, 56.— NAVIGATION; THE SAILOR'S SEA-BOOK.—How to Keep the Log and Work it Off—Latitude and Longitude—Great Circle Sailing—Law of Storms and variable Winds; and an Explanation of Terms used, with coloured Illustrations of Flags.

John Weale, 59, High Holborn, London, W.C.

MR. WEALE'S RUDIMENTARY SERIES.

In demy 12mo, with Woodcuts, cloth, price 2s.

RUDIMENTARY.—57, 58.—WARMING AND VENTILATION, by CHARLES TOMLINSON, 2 vols. in 1.

In demy 12mo, with Woodcuts, cloth, price 1s.

RUDIMENTARY.—59.—STEAM BOILERS, by R. ARMSTRONG, C.E.

In demy 12mo, with Woodcuts, cloth, price 2s.

RUDIMENTARY. — 60, 61. — LAND AND ENGINEERING SURVEYING, by T. BAKER, C.E., 2 vols. in 1.

In demy 12mo, with Woodcuts, cloth, price 1s.

RUDIMENTARY AND ELEMENTARY.—62. —PRINCIPLES OF RAILWAYS, for the Use of the Beginner in his Studies; with Sketches for Construction. By Sir R. MACDONALD STEPHENSON. Vol. I.

In demy 12mo, with Woodcuts, cloth, price 1s.

RUDIMENTARY.—62*.—RAILWAY WORKING IN GREAT BRITAIN, Statistical Details, Table of Capital and Dividends, Revenue Accounts, Signals, &c., Vol. II.

In demy 12mo, with Woodcuts, cloth, price 3s.

RUDIMENTARY.—63, 64, 65.—AGRICULTURAL BUILDINGS, the Construction of, on Motive Powers, and the Machinery of the Steading; and on Agricultural Field Engines, Machines, and Implements, by G. H. ANDREWS, 3 vols in 1. -John Weale, 59, High Holborn, London, W.C.

In demy 12mo, cloth, price 1s.

RUDIMENTARY.—66.—CLAY LANDS AND LOAMY SOILS, by Professor JOHN DONALDSON, A.E.

In demy 12mo, with Woodcuts, cloth, price 3s.

RUDIMENTARY. — 67, 68. — CLOCK AND WATCH-MAKING, AND ON CHURCH CLOCKS AND BELLS, by E. B. DENISON, M.A., 2 vols. in 1, considerably extended. Fourth Edition.

In demy 12mo, with Woodcuts, cloth, price 2s.

RUDIMENTARY.—69, 70.— MUSIC, Practical Treatise on, by C. C. SPENCER, Mus. Dr. 2 vols. in 1.

In demy 12mo, cloth, price 1s.

RUDIMENTARY.— 71.—PIANOFORTE, Instruction for Playing the, by C. C. SPENCER, Mus. Dr.

In demy 12mo, with Steel Engravings and Woodcuts, cloth, price 5s. 6d.

RUDIMENTARY.—72, 73, 74, 75, 75*.—RECENT FOSSIL SHELLS (A Manual of the Mollusca), by SAMUEL P. WOODWARD, of the Brit. Mus. 4 vols. in 1, with Supplement.

In demy 12mo., with Woodcuts, cloth, price 2s.

RUDIMENTARY. — 76, 77. — DESCRIPTIVE GEOMETRY, by J. F. HEATHER, M.A. 2 vols. in 1.

In demy 12mo, with Woodcuts, price 1s.

RUDIMENTARY. — 77*. — ECONOMY OF FUEL, by T. S. PRIDEAUX.

In demy 12mo, 2 vols. in 1, with Woodcuts, cloth, price 2s.

RUDIMENTARY.—78, 79.—STEAM AS APPLIED TO GENERAL PURPOSES.

John Weale, 59, High Holborn, London, W.C.

MR. WEALE'S RUDIMENTARY SERIES.

In demy 12mo, with Woodcuts, cloth, price 1s. 6d.

RUDIMENTARY.—78*.—LOCOMOTIVE ENGINE, by G. D. DEMPSEY, C.E.

In royal 4to, cloth, price 4s. 6d.

RUDIMENTARY. — 79*. — ATLAS OF ENGRAVED PLATES to DEMPSEY'S LOCOMOTIVE ENGINES.

In demy 12mo, with Woodcuts, cloth, price 1s.

RUDIMENTARY.—79**.—ON PHOTOGRAPHY, the Composition and Properties of the Chemical Substances used, by Dr. H. HALLEUR.

In demy 12mo., with Woodcuts, cloth, price 2s. 6d.

RUDIMENTARY. — 80, 81. — MARINE ENGINES AND ON THE SCREW, &c., by R. MURRAY, C.E. 2 vols. in 1.

In demy 12mo, cloth, price 2s.

RUDIMENTARY.—80*, 81*.—EMBANKING LANDS FROM THE SEA, by JOHN WIGGINS, F.G.S. 2 vols. in 1.

In demy 12mo, with Woodcuts, cloth, price 2s.

RUDIMENTARY. — 82, 82*. — POWER OF WATER, AS APPLIED TO DRIVE FLOUR MILLS, by JOSEPH GLYNN, F.R.S., C.E.

In demy 12mo, cloth, price 1s.

RUDIMENTARY.—83.—BOOK-KEEPING, by JAMES HADDON, M.A.

In demy 12mo, with Woodcuts, price 3s.

RUDIMENTARY. — 82**, 83*, 83 (*bis*) COAL GAS, on the Manufacture and Distribution of, by SAMUEL HUGHES, C.E.

In demy 12mo, with Woodcuts, cloth, price 3s.

RUDIMENTARY.—82***.—WATER WORKS FOR THE SUPPLY OF CITIES AND TOWNS; Works which have been executed for procuring Supplies by means of Drainage Areas and by Pumping from Wells, by SAMUEL HUGHES, C.E.

In demy 12mo, with Woodcuts, cloth, price 1s. 6d.

RUDIMENTARY.—83**.—CONSTRUCTION OF DOOR LOCKS.

In demy 12mo, with Woodcuts, cloth, price 1s.

RUDIMENTARY. — 83 (*bis*) — FORMS OF SHIPS AND BOATS, by W. BLAND, of Hartlip.

In demy 12mo, cloth, price 1s. 6d.

RUDIMENTARY.—84.—ARITHMETIC, with numerous Examples, by Prof. J. R. YOUNG.

In demy 12mo, cloth, price 1s. 6d.

RUDIMENTARY. —84*.— KEY to the above, by Prof. J. R. YOUNG.

In demy 12mo, cloth, price 1s.

RUDIMENTARY. — 85. — EQUATIONAL ARITHMETIC, Questions of Interest, Annuities, &c., by W. HIPSLEY.

John Weale, 59, High Holborn, London, W.C.

MR. WEALE'S RUDIMENTARY SERIES.

In demy 12mo, cloth, price 1s.

RUDIMENTARY.—85*.—SUPPLEMENTARY VOLUME TO HIPSLEY'S EQUATIONAL ARITHMETIC, Tables for the Calculation of Simple Interest, with Logarithms for Compound Interest and Annuities, &c., &c., by W. HIPSLEY.

In demy 12mo, cloth, price 2s.

RUDIMENTARY.—86, 87.—ALGEBRA, by JAMES HADDON, M.A. 2 vols. in 1.

In demy 12mo, in cloth, price 1s. 6d.

RUDIMENTARY.—86*, 87*.—ELEMENTS OF ALGEBRA, Key to the, by Prof. YOUNG.

In demy 12mo, with Woodcuts, price 2s.

RUDIMENTARY.—88, 89.—ELEMENTS OF GEOMETRY, by HENRY LAW, C.E. 2 vols. in 1.

In demy 12mo, with Woodcuts, cloth, price 1s.

RUDIMENTARY.—90.—GEOMETRY, ANALYTICAL, by Prof. JAMES HANN.

In demy 12mo, with Woodcuts, cloth, price 2s.

RUDIMENTARY.—91, 92.—PLANE AND SPHERICAL TRIGONOMETRY, by the same. 2 vols. in 1.

In demy 12mo, with Woodcuts, cloth, price 1s.

RUDIMENTARY.—93.—MENSURATION, by T. BAKER, C.E.

In demy 12mo, cloth, price 2s. 6d.

RUDIMENTARY.—94, 95.—LOGARITHMS, Tables for facilitating Astronomical, Nautical, Trigonometrical, and Logarithmic Calculations, by H. LAW, C.E. New Edition, with Tables of Natural Sines and Tangents, and Natural Cosines. 2 vols. in 1.

In demy 12mo, with Woodcuts, cloth, price 1s.

RUDIMENTARY.—96.—POPULAR ASTRONOMY. By the Rev. ROBERT MAIN, M.R.A.S.

In demy 12mo, with Woodcuts, cloth, price 1s.

RUDIMENTARY.—97.—STATICS AND DYNAMICS, by T. BAKER, C.E.

In demy 12mo, with 220 Woodcuts, cloth, price 2s. 6d.

RUDIMENTARY.—98, 98*.—MECHANISM AND PRACTICAL CONSTRUCTION OF MACHINES, by T. BAKER, C.E., and ON TOOLS AND MACHINES, by JAMES NASMYTH, C.E.

In demy 12mo, with Woodcuts, cloth, price 2s.

RUDIMENTARY.—99, 100.—NAUTICAL ASTRONOMY AND NAVIGATION, by Prof. YOUNG. 2 vols. in 1.

In demy 12mo, cloth, price 1s. 6d.

RUDIMENTARY.—100*.—NAVIGATION TABLES, compiled for practical use with the above.

In demy 12mo, cloth, price 1s.

RUDIMENTARY.—101.—DIFFERENTIAL CALCULUS, by Mr. WOOLHOUSE, F.R.A.S.

MR. WEALE'S RUDIMENTARY SERIES.

In demy 12mo, cloth, price 1s. 6d.

RUDIMENTARY. — 101*. — WEIGHTS AND MEASURES OF ALL NATIONS: Weights, Coins, and the various Divisions of Time, with the principles which determine Rates of Exchange, by Mr. WOOLHOUSE, F.R.A.S.

In demy 12mo, in cloth, price 1s.

RUDIMENTARY. — 102. — INTEGRAL CALCULUS, by H. COX, M.A.

In demy 12mo, in cloth, price 1s.

RUDIMENTARY. — 103. — INTEGRAL CALCULUS, Examples of, by Prof. JAMES HANN.

In demy 12mo, cloth, price 1s.

RUDIMENTARY. — 104. — DIFFERENTIAL CALCULUS, Examples of, by J. HADDON, M.A.

In demy 12mo, with Woodcuts, cloth, price 1s. 6d.

RUDIMENTARY. — 105. — ALGEBRA, GEOMETRY, AND TRIGONOMETRY, Mnemonical Lessons, by the Rev. T. PENYNGTON KIRKMAN, M.A.

In demy 12mo, with Woodcuts, cloth, price 1s. 6d.

RUDIMENTARY. — 106. — SHIPS' ANCHORS FOR ALL SERVICES, by Mr. GEORGE COTSELL, N.A.

In demy 12mo, with Woodcuts, price 2s. 6d.

RUDIMENTARY. — 107. — METROPOLITAN BUILDINGS ACT in present operation, with Notes, and the Act dated August 28th, 1860, for better supplying of Gas to the Metropolis.

In demy 12mo, cloth, price 1s. 6d.

RUDIMENTARY. — 108. — METROPOLITAN LOCAL MANAGEMENT ACTS. All the Acts.

In demy 12mo, cloth, price 1s. 6d.

RUDIMENTARY. — 109. — LIMITED LIABILITY AND PARTNERSHIP ACTS.

In demy 12mo, cloth, price 1s.

RUDIMENTARY. — 110. — SIX RECENT LEGISLATIVE ENACTMENTS, for Contractors, Merchants, and Tradesmen.

In demy 12mo, cloth, price 1s.

RUDIMENTARY. — 111. — NUISANCES REMOVAL AND DISEASE PREVENTION ACT.

In demy 12mo, cloth, price 1s. 6d.

RUDIMENTARY. — 112. — DOMESTIC MEDICINE, PRESERVING HEALTH, by M. RASPAIL.

In demy 12mo, cloth, price 1s. 6d.

RUDIMENTARY. — 113. — USE OF FIELD ARTILLERY ON SERVICE, by Lieut.-Col. HAMILTON MAXWELL, B.A.

In demy 12mo, with Woodcuts, cloth, price 1s. 6d.

RUDIMENTARY. — 114. — ON MACHINERY: Rudimentary and Elementary Principles of the Construction and on the Working of Machinery, by C. D. ABEL, C.E.

In royal 4to, cloth, price 7s. 6d.

RUDIMENTARY. — 115. — ATLAS OF PLATES OF SEVERAL KINDS OF MACHINES, 17 very valuable Illustrative plates.

John Weale, 59, High Holborn, London, W.C.

MR. WEALE'S RUDIMENTARY SERIES.

In demy 12mo, with Woodcuts, cloth, price 1s. 6d.

RUDIMENTARY. — 116. — TREATISE ON ACOUSTICS: The Distribution of Sound, by T. ROGER SMITH, Architect.

In demy 12mo, with Woodcuts, cloth, price 2s. 6d.

RUDIMENTARY.—117.—SUBTERRANEOUS SURVEYING, RANGING THE LINE WITHOUT THE MAGNET. By THOMAS FENWICK, Coal Viewer. With Improvements and Modern Additions by T. BAKER, C.E.

In demy 12mo, with Plates and Woodcuts, cloth, price 3s.

RUDIMENTARY.—118, 119.—ON THE CIVIL ENGINEERING OF NORTH AMERICA, by D. STEVENSON, C.E. 2 vols. in 1.

In demy 12mo, with Woodcuts, cloth, price 3s.

RUDIMENTARY. — 120. — ON HYDRAULIC ENGINEERING, by G. R. BURNELL, C.E. 2 vols. in 1.

In demy 12mo, with 2 Engraved Plates, cloth, price 1s. 6d.

RUDIMENTARY. — 121. — TREATISE ON RIVERS AND TORRENTS, from the Italian of PAUL FRISI.

In demy 12mo, by PAUL FRISI, in cloth, price 1s.

RUDIMENTARY.—122.—ON RIVERS THAT CARRY SAND AND MUD, and an ESSAY ON NAVIGABLE CANALS. 121 and 122 bound together, 2s. 6d.

In demy 12mo, with Woodcuts, cloth, price 1s. 6d.

RUDIMENTARY. — 123. — ON CARPENTRY AND JOINERY, founded on Dr. Robison's Work.

In demy 4to, cloth, price 4s. 6d.

RUDIMENTARY.—123*.—ATLAS of PLATES in detail to the CARPENTRY AND JOINERY. 123 and 123* bound together in cloth in 1 vol.

In demy 12mo, with Woodcuts, cloth, price 1s. 6d.

RUDIMENTARY. — 124. — ON ROOFS FOR PUBLIC AND PRIVATE BUILDINGS, founded on Dr. Robison's Work.

In royal 4to, cloth, price 4s. 6d.

RUDIMENTARY.—124*.—RECENTLY CONSTRUCTED IRON ROOFS. Atlas of plates.

In demy 12mo, with Woodcuts, cloth, price 3s.

RUDIMENTARY.—125.—ON THE COMBUSTION OF COAL AND THE PREVENTION OF SMOKE, Chemically and Practically Considered, by CHARLES WYE WILLIAMS.

In demy 12mo, cloth. 125 and 126 together, price 3s.

RUDIMENTARY. — 126. — ILLUSTRATIONS to WILLIAMS'S COMBUSTION OF COAL. 125 and 126, 2 vols. bound in 1.

In demy 12mo, with Woodcuts, cloth, price 1s. 6d.

RUDIMENTARY. — 127. — PRACTICAL INSTRUCTIONS IN THE ART OF ARCHITECTURAL MODELLING.

John Weale, 59, High Holborn, London, W.C.

MR. WEALE'S RUDIMENTARY SERIES.

In demy 12mo, with Engravings and Woodcuts.

RUDIMENTARY.—128.—THE TEN BOOKS OF M. VITRUVIUS ON CIVIL, MILITARY, AND NAVAL ARCHITECTURE, translated by JOSEPH GWILT, Arch. 2 vols. in 1.

In demy 12mo, 128 and 129 together, cloth, price 5s.

RUDIMENTARY. — 129. — ILLUSTRATIVE PLATES TO VITRUVIUS'S TEN BOOKS, by the Author and JOSEPH GANDY, R.A.

In demy 12mo, cloth, price 1s.

RUDIMENTARY. — 130. — INQUIRY INTO THE PRINCIPLES OF BEAUTY IN GRECIAN ARCHITECTURE, by the Right Hon. the Earl of ABERDEEN, &c. &c.

In demy 12mo, cloth, price 1s.

RUDIMENTARY. — 131. — THE MILLER'S, MERCHANT'S, AND FARMER'S READY RECKONER, for ascertaining at Sight the Value of any quantity of Corn; together with the approximate value of Millstones and Millwork.

In demy 12mo, with Woodcuts, cloth, price 2s. 6d.

RUDIMENTARY.—132.—TREATISE ON THE ERECTION OF DWELLING HOUSES, WITH SPECIFICATIONS, QUANTITIES OF THE VARIOUS MATERIALS, &c., by S. H. BROOKS, Architect. 27 Plates.

RUDIMENTARY SERIES. — ON MINES, SMELTING WORKS, AND THE MANUFACTURE OF METALS, as follows.

In demy 12mo, with Woodcuts, cloth, price 2s.

RUDIMENTARY. — Vol. 1. — TREATISE ON THE METALLURGY OF COPPER, by R. H. LAMBORN.

In demy 12mo, to have Woodcuts, cloth.

RUDIMENTARY. — Vol. 2. — TREATISE ON THE METALLURGY OF SILVER AND LEAD.

In demy 12mo, to have Woodcuts, cloth.

RUDIMENTARY AND ELEMENTARY. — Vol. 3.—TREATISE ON IRON METALLURGY up to the Manufacture of the latest processes.

In demy 12mo, to have Woodcuts, cloth.

RUDIMENTARY AND ELEMENTARY. — Vol. 4.—TREATISE ON GOLD MINING AND ASSAYING PLATINUM, IRIDIUM, &c.

In demy 12mo, to have Woodcuts, cloth.

RUDIMENTARY AND ELEMENTARY. — Vol. 5.—TREATISE ON THE MINING OF ZINC, TIN, NICKEL, COBALT, &c.

In demy 12mo, to have Woodcuts, cloth.

RUDIMENTARY AND ELEMENTARY. — Vol. 6.—TREATISE ON COAL MINING (Geology and Means of Discovering, &c.)

In demy 12mo, with Woodcuts, cloth, price 1s. 6d.

RUDIMENTARY. — Vol. 7. — ELECTRO-METALLURGY. — Practically treated by ALEXANDER WATT, F.R.S.A.

John Weale, 59, High Holborn, London, W.C.

NEW SERIES OF EDUCATIONAL WORKS.

In demy 12mo, with Woodcuts, cloth, price 4s.

CONSTITUTIONAL HISTORY OF ENGLAND.—1, 2, 3, 4.—By W. D. HAMILTON, of the State P. O.

In demy 12mo, with Woodcuts, cloth, price 2s. 6d.

OUTLINES OF THE HISTORY OF GREECE.—5, 6.—By W. D. HAMILTON, 2 vols.

In demy 12mo, with Map of Italy and Woodcuts, cloth, price 2s. 6d

OUTLINE OF THE HISTORY OF ROME.—7, 8.—By W. D. HAMILTON, 2 vols.

In demy 12mo, cloth, price 2s. 6d.

CHRONOLOGY OF CIVIL AND ECCLESIASTICAL HISTORY, LITERATURE, ART, AND CIVILISATION, from the earliest period to the present.—9, 10.—2 vols.

In demy 12mo, cloth, price 1s.

GRAMMAR OF THE ENGLISH LANGUAGE.—11.—By HYDE CLARKE, D.C.L.

In demy 12mo, cloth, price 1s.

HANDBOOK OF COMPARATIVE PHILOLOGY.—11*.—By HYDE CLARKE, D.C.L.

In demy stout 12mo, cloth, price 3s. 6d.

DICTIONARY OF THE ENGLISH LANGUAGE.—12, 13.—A New Dictionary of the English Tongue as spoken and written, above 100,000 words, or 50,000 more than in any existing work, by HYDE CLARKE, D.C.L., 3 vols. in 1.

In demy 12mo, cloth, price 1s.

GRAMMAR OF THE GREEK LANGUAGE.—14.—By H. C. HAMILTON.

In demy 12mo, cloth, price 2s.

DICTIONARY OF THE GREEK AND ENGLISH LANGUAGES.—15, 16.—By H. R. HAMILTON, 2 vols. in 1.

In demy 12mo, cloth, price 2s.

DICTIONARY OF THE ENGLISH AND GREEK LANGUAGES.—17, 18.—By H. R. HAMILTON, 2 vols. in 1.

In demy 12mo, cloth, price 1s.

GRAMMAR OF THE LATIN LANGUAGE.—19.—By the Rev. T. GOODWIN, A.B.

In demy 12mo, cloth, price 2s.

DICTIONARY OF THE LATIN AND ENGLISH LANGUAGES.—20, 21.—By the Rev. T. GOODWIN, B.A. Vol. I.

In demy 12mo, cloth, price 1s. 6d.

DICTIONARY OF THE ENGLISH AND LATIN LANGUAGES.—22, 23.—By the Rev. T. GOODWIN, A.B. Vol. II.

In demy 12mo, cloth, price 1s.

GRAMMAR OF THE FRENCH LANGUAGE.—24.

MR. WEALE'S EDUCATIONAL SERIES.

In demy 12mo, cloth, price 1s.

DICTIONARY OF THE FRENCH AND ENGLISH LANGUAGES.—25.—By A. ELWES. Vol. I.

In demy 12mo, cloth, price 1s. 6d.

DICTIONARY OF THE ENGLISH AND FRENCH LANGUAGES.—26.—By A. ELWES. Vol. II.

In demy 12mo, cloth, price 1s.

GRAMMAR OF THE ITALIAN LANGUAGE—27.—By A. ELWES.

In demy 12mo, cloth, price 2s.

DICTIONARY OF THE ITALIAN, ENGLISH, AND FRENCH LANGUAGES.—28, 29.—By A. ELWES. Vol. I.

In demy 12mo, cloth, price 2s.

DICTIONARY OF THE ENGLISH, ITALIAN, AND FRENCH LANGUAGES.—30, 31.—By A. ELWES. Vol. II.

In demy 12mo, cloth, price 2s.

DICTIONARY OF THE FRENCH, ITALIAN, AND ENGLISH LANGUAGES.—32, 33.—By A. ELWES. Vol. III.

In demy 12mo, cloth, price 1s.

GRAMMAR OF THE SPANISH LANGUAGE.—34.—By A. ELWES.

In demy 12mo, cloth, price 4s.

DICTIONARY OF THE SPANISH AND ENGLISH LANGUAGES.—35, 36, 37, 38.—By A. ELWES. 4 vols. in 1.

In demy 12mo, cloth, price 1s.

GRAMMAR OF THE GERMAN LANGUAGE.—39.

In demy 12mo, cloth, price 1s.

CLASSICAL GERMAN READER.—40.—From the best Authors.

In demy 12mo, cloth, price 3s.

DICTIONARIES OF THE ENGLISH, GERMAN, AND FRENCH LANGUAGES.—41, 42, 43.—By N. E. HAMILTON, 3 vols., separately, 1s. each.

In demy 12mo, cloth, price 7s.

DICTIONARY OF THE HEBREW AND ENGLISH LANGUAGES.—44, 45.—Containing the Biblical and Rabbinical words, 2 vols. (together with the Grammar, which may be had separately for 1s.), by Dr. BRESSLAU, Hebrew Professor.

In demy 12mo, cloth, price 3s.

DICTIONARY OF THE ENGLISH AND HEBREW LANGUAGES.—46.—Vol. III. to complete.

In demy 12mo, cloth, price 1s.

FRENCH AND ENGLISH PHRASE BOOK.—47.

John Weale, 59, High Holborn, London, W.C.

MR. WEALE'S CLASSICAL SERIES.

Now in course of Publication, in demy 12mo, price 1s. per Volume (except in some instances, and those are 1s. 6d. or 2s. each), very neatly printed on good paper. Those priced are published.

GREEK AND LATIN CLASSICS.—A Series of Volumes containing the principal Greek and Latin Authors, accompanied by Explanatory Notes in English, principally selected from the best and most recent German Commentators, and comprising all those Works that are essential for the Scholar and the Pupil, and applicable for the Universities of Oxford, Cambridge, Edinburgh, Glasgow, Aberdeen, and Dublin—the Colleges at Belfast, Cork, Galway, Winchester, and Eton, and the great Schools at Harrow, Rugby, &c.—also for Private Tuition and Instruction, and for the Library, as follows:

LATIN SERIES.

In demy 12mo, boards, price 1s.

A NEW LATIN DELECTUS.—1.—Extracts from Classical Authors, with Vocabularies and Explanatory Notes.

In demy 12mo, boards, price 2s.

CÆSAR'S COMMENTARIES ON THE GALLIC WAR.—2.—With Grammatical and Explanatory Notes in English, and a Geographical Index.

In demy 12mo, boards, price 1s.

CORNELIUS NEPOS.—3.—With English Notes, &c.

In demy 12mo, boards, price 1s.

VIRGIL.—4.—The Georgics, Bucolics, with English Notes.

In demy 12mo, boards, price 2s.

VIRGIL'S ÆNEID.—5.—(On the same plan as the preceding).

In demy 12mo, boards, price 1s.

HORACE.—6.—Odes and Epodes; with English Notes, and Analysis and Explanation of the Metres.

In demy 12mo, boards, price 1s. 6d.

HORACE.—7.—Satires and Epistles, with English Notes, &c.

In demy 12mo, boards, price 1s. 6d.

SALLUST.—8.—Conspiracy of Catiline, Jugurthine War, with English Notes.

In demy 12mo, boards, price 1s. 6d.

TERENCE.—9.—Andrea and Heautontimorumenos, with English Notes.

In demy 12mo, boards, price 2s.

TERENCE.—10.—Phormio, Adelphi, and Hecyra, with English Notes.

In demy 12mo.

CICERO.—11.—Orations against Catiline, for Sulla, for Archias, and for the Manilian Law.

In demy 12mo.

CICERO.—12.—First and Second Philippics; Orations for Milo, for Marcellus, &c.

John Weale, 59, High Holborn London, W.C.

MR. WEALE'S CLASSICAL SERIES.

In demy 12mo.

CICERO.—13.—De Officiis.

In demy 12mo, boards, price 2s.

CICERO.—14.—De Amicitiâ, de Senectute, and Brutus, with English Notes.

In demy 12mo.

JUVENAL AND PERSIUS.—15.—(The indelicate parts expunged.)

In demy 12mo, boards, price 3s.

LIVY. — 16. — Books i. to v. in two vols., with English Notes.

In demy 12mo, boards, price 1s.

LIVY.—17.—Books xxi. and xxii., with English Notes.

In demy 12mo.

TACITUS.—18.—Agricola; Germania; and Annals, Book I.

In demy 12mo, boards, price 2s.

SELECTIONS FROM TIBULLUS, OVID, and PROPERTIUS.—19.—With English Notes.

In demy 12mo.

SELECTIONS FROM SUETONIUS and the later Latin Writers.—20.

GREEK SERIES, ON A SIMILAR PLAN TO THE LATIN SERIES.

Those not priced are in the Press.

In demy 12mo, boards, price 1s.

INTRODUCTORY GREEK READER. — 1. — On the same plan as the Latin Reader.

In demy 12mo, boards, price 1s.

XENOPHON. — 2. — Anabasis, i. ii. iii., with English Notes.

In demy 12mo, boards, price 1s.

XENOPHON. — 3. — Anabasis, iv. v. vi. vii., with English Notes.

In demy 12mo, boards, price 1s.

LUCIAN. — 4. — Select Dialogues, with English Notes.

In demy 12mo, boards, price 1s. 6d.

HOMER.—5.—Iliad, i. to vi., with English Notes.

In demy 12mo, boards, price 1s. 6d.

HOMER.—6.—Iliad, vii. to xii., with English Notes.

In demy 12mo, boards, price 1s. 6d.

HOMER. — 7. — Iliad, xiii. to xviii. with English Notes.

In demy 12mo. boards, price 1s 6d.

HOMER. — 8. — Iliad, xix. to xxiv., with English Notes.

MR. WEALE'S CLASSICAL SERIES.

In demy 12mo, boards, price 1s. 6d.

HOMER.—9.—Odyssey, i. to vi., with English Notes.

In demy 12mo, boards, price 1s. 6d.

HOMER.—10.—Odyssey, vii. to xii., with English Notes.

In demy 12mo, boards, price 1s. 6d.

HOMER.—11.—Odyssey, xiii. to xviii. with English Notes.

In demy 12mo, boards, price 1s. 6d.

HOMER. — 12. — Odyssey, xix. to xxiv.; and Hymns, with English Notes.

In demy 12mo, boards, price 2s.

PLATO. — 13. — Apology, Crito, and Phædo, with English Notes.

In demy 12mo, boards, price 1s. 6d.

HERODOTUS.—14.—i. ii., with English Notes.—Dedicated to His Grace the Duke of Devonshire.

In demy 12mo, boards, price 1s. 6d.

HERODOTUS.—15.—iii. iv., with English Notes. Dedicated to His Grace the Duke of Devonshire.

In demy 12mo.

HERODOTUS.—16.—v. vi. and part of vii. Dedicated to His Grace the Duke of Devonshire.

In demy 12mo.

HERODOTUS.—17.—Remainder of vii., viii., and ix. Dedicated to His Grace the Duke of Devonshire.

In demy 12mo, boards, price 1s.

SOPHOCLES. — 18. — Œdipus Rex, with English Notes.

In demy 12mo.

SOPHOCLES.—19.—Œdipus Colonæus.

In demy 12mo.

SOPHOCLES.—20.—Antigone.

In demy 12mo.

SOPHOCLES.—21.—Ajax.

In demy 12mo.

SOPHOCLES.—22.—Philoctetes.

In demy 12mo, boards, price 1s. 6d.

EURIPIDES.—23.—Hecuba, with English Notes.

In demy 12mo.

EURIPIDES.—24.—Medea.

In demy 12mo.

EURIPIDES.—25.—Hippolytus.

MR. WEALE'S CLASSICAL SERIES.

In demy 12mo, boards, price 1s.
EURIPIDES.—26.—Alcestis, with English Notes.

In demy 12mo.
EURIPIDES.—27.—Orestes.

In demy 12mo.
EURIPIDES.—28.—Extracts from the remaining Plays.

In demy 12mo.
SOPHOCLES.—29.—Extracts from the remaining Plays.

In demy 12mo.
ÆSCHYLUS.—30.—Prometheus Vinctus.

In demy 12mo.
ÆSCHYLUS.—31.—Persæ.

In demy 12mo.
ÆSCHYLUS.—32.—Septem contra Thebas.

In demy 12mo.
ÆSCHYLUS.—33.—Choëphoræ.

In demy 12mo.
ÆSCHYLUS.—34.—Eumenides.

In demy 12mo.
ÆSCHYLUS.—35.—Agamemnon.

In demy 12mo.
ÆSCHYLUS.—36.—Supplices.

In demy 12mo.
PLUTARCH.—37.—Select Lives.

In demy 12mo.
ARISTOPHANES.—38.—Clouds.

In demy 12mo.
ARISTOPHANES.—39.—Frogs.

In demy 12mo.
ARISTOPHANES.—40.—Selections from the remaining Comedies.

In demy 12mo, boards, price 1s.
THUCYDIDES.—41.—I., with English Notes.

In demy 12mo.
THUCYDIDES.—42.—II.

MR. WEALE'S CLASSICAL SERIES.

In demy 12mo.

THEOCRITUS.—43.—Select Idyls.

In demy 12mo.

PINDAR.—44.

In demy 12mo.

SOCRATES.—45.

In demy 12mo.

HESIOD.—46.

MR. WEALE'S PUBLICATIONS OF WORKS ON ARCHITECTURE, ENGINEERING, AND THE FINE ARTS.

In 1 large Atlas, folio Volume, with fine Plates, price £4 4s.

"BRITISH GOVERNMENT WORK."—THE ARCHITECTURAL ANTIQUITIES AND RESTORATION OF ST. STEPHEN'S CHAPEL, WESTMINSTER (late the House of Commons).

Fine Plates and Vignettes, Atlas folio, price £3 10s.

"NORWEGIAN GOVERNMENT WORK." —THE CATHEDRAL OF THRONDHEIM, IN NORWAY. Text by Professor MUNCH; drawings by H. E. SCHIRMER, Architect.

Large Atlas folio, 4 livraisons, published in Madrid, at 100 reals each, or £1 in England. Illustrated by beautifully executed Engravings, some of which are coloured.

"SPANISH GOVERNMENT WORK."—MONUMENTS ARCHITECTONIQUES DE L'ESPAGNE, PUBLIÉS AUX FRAIS DE LA NATION.—PART I Provincia de Toledo, Granada, Alcalá de Henares.—PART 2. Catedral Toledo, Detallies.—PART 3. Granada, Segovia, Toledo, Salamanca.—PART 4. Santa Maria de Alcalá de Henares, Casa Lonia de Valencia, Toledo, Segovia, &c.—This work surpasses in beauty all other works.

Columbier folio plates, with text also uniform, with gold borders, and sumptuously bound in red morocco, gilt; gilt leaves, £12 12s., Columbier folio plates, with text also uniform, with gold borders, and elegantly half-bound in morocco, gilt, £10 10s.; Plates in Columbier folio, and text in Imperial 4to, half-bound in morocco, gilt, £7 7s.; Plates in Columbier folio, and text in imperial 4to, in cloth extra, boards and lettered, £4 14s. 6d.

THE VICTORIA BRIDGE, AT MONTREAL, IN CANADA. — Elaborately illustrated by views, plans, elevations, and details of the Bridge; together with the illustrations of the Machinery and Contrivances used in the construction of this stupendously important and valuable engineering work. The whole produced in the finest style of art, pictorially and geometrically drawn, and the views highly coloured, and a descriptive text. Dedicated to His Royal Highness the Prince of Wales. By JAMES HODGES, Engineer to the Contractors. Engineers: ROBERT STEPHENSON and ALEX. M. ROSS. Contractors: Sir S. MORTON PETO, Bart., M.P., THOMAS BRASSEY, and EDWARD LADD BETTS, Esqrs.

John Weale, 59, High Holborn, London, W.C.

MR. WEALE'S WORKS ON ARCHITECTURE, ENGINEERING, FINE ARTS, &c.

In 4to, 1s. 6d.

ARAGO, Mons. — Report on the Atmospheric System, and on the proposed Atmospheric Railway at Paris.

In 4to, with about 500 Engravings, some of which are highly coloured, 4 vols., original copies, half-bound in morocco, £6 6s.

ARCHITECTURAL PAPERS.

2 Engravings, in folio, useful to learners and for schools, 2s. 6d.

ARCHITECTURAL ORDERS (FIVE) AND THEIR ENTABLATURES, drawn to a larger scale, with Figured Dimensions.

4to, 1s.

ARNOLLET, M. — Report on his Atmospheric Railway.

In 4to, 10 Plates, 7s. 6d.

ATMOSPHERIC RAILWAYS. — THREE REPORTS on improved methods of Constructing and Working Atmospheric Railways. By R. MALLET, C.E.

In 8vo, 1s. 6d.

BARLOW, P. W. —Observations on the Niagara Railway Suspension Bridge.

In large 4to, very neat half-morocco, 18s., with Engravings.

BARRY, SIR CHARLES, R.A., &c. — Studies of Modern English Architecture. By W. H. LEEDS; The Travellers' Club-House, illustrated by Engravings of Plans, Sections, Elevations, and details.

In 1 Vol., large 8vo, with coloured Plates, half-morocco, price £1 1s.

BEWICK'S (J. G.) GEOLOGICAL TREATISE ON THE DISTRICT OF CLEVELAND IN NORTH YORKSHIRE, its Ferruginous Deposits, Lias and Oolites; with some Observations on Ironstone Mining.

In 8vo, with Plates. Price 4s.

BINNS, W. S. — Work on Geometrical Drawing, embracing Practical Geometry, including the use of Drawing Instruments, the construction and use of Scales, Orthographic Projection, and Elementary Descriptive Geometry.

In 4to, with 106 Illustrative Plates, cloth boards, £1 11s. 6d.

BLASHFIELD, J. M., M. R. Inst., &c.— SELECTIONS OF VASES, STATUES, BUSTS, &c, from TERRA COTTAS.

In 8vo, Woodcuts, 1s.

BLASHFIELD, J. M., M. R., Inst., &c.— ACCOUNT OF THE HISTORY AND MANUFACTURE OF ANCIENT AND MODERN TERRA COTTA.

In 4to, 2s. 6d.

BODMER, R., C.E.—On the Propulsion of Vessels by the Screw.

15s.

BRIDGE. — A large magnificent Plate, 3 feet 6 inches by 2 feet, on a scale of 25 feet to an inch, of LONDON BRIDGE; containing Plan and Elevation. Engraved and elaborately finished. The Work of the RENNIES.

John Weale, 59, High Holborn, London, W.C.

MR. WEALE'S WORKS ON ARCHITECTURE, ENGINEERING, FINE ARTS, &c.

10s.

BRIDGE. — Plan and Elevation, on a scale of 10 feet to an inch, of STAINES BRIDGE; a fine Engraving. The work of the RENNIES.

In royal 8vo, with very elaborate Plates (folded), £1 10s.

BRIDGES, SUSPENSION. — An Account, with Illustrations, of the Suspension Bridge across the River Danube, by Wm. T. CLARK, F.R.S.

In 4 vols., royal 8vo, bound in 3 vols., half-morocco, price £4 10s.

BRIDGES. — THE THEORY, PRACTICE, AND ARCHITECTURE OF BRIDGES OF STONE, IRON, TIMBER, AND WIRE; with Examples on the Principle of Suspension; Illustrated by 138 Engravings and 92 Woodcuts.

In one large 8vo volume, with explanatory Text, and 68 Plates comprising details and measured dimensions. Bound in half-morocco, uniform with the preceding work, price £2 10s.

BRIDGES. — SUPPLEMENT TO "THE THEORY, PRACTICE, AND ARCHITECTURE OF BRIDGES OF STONE, IRON, TIMBER, WIRE, AND SUSPENSION."

1 large folio Engraving, price 7s. 6d.

BRIDGE across the Thames.—SOUTHWARK IRON BRIDGE.

1 large folio Engraving, price 5s.

BRIDGE across the Thames. — WATERLOO STONE BRIDGE.

1 very large Engraving, price 5s.

BRIDGE across the Thames. — VAUXHALL IRON BRIDGE.

1 very large Engraving, price 4s. 6d.

BRIDGE across the Thames.—HAMMERSMITH SUSPENSION BRIDGE.

1 large Engraving, price 4s. 6d.

BRIDGE (the UPPER SCHUYLKILL) at PHILADELPHIA, the greatest known span of one arch, covered.

1 large Engraving, price 3s. 6d.

BRIDGE (the SCHUYLKILL) at PHILADELPHIA, covered.

1 large Engraving, price 3s. 6d.

BRIDGE. — ON THE PRINCIPLE OF SUSPENSION, by Sir I. BRUNEL, in the ISLAND OF BOURBON.

1 large Engraving, price 4s.

BRIDGE. — PLAN and ELEVATION of the PATENT IRON BAR BRIDGE over the River Tweed, near Berwick.

34 Plates, folio, £1 1s., boards.

BRIGDEN, R. — Interior Decorations, Details, and Views of Sefton Church, Lancashire, erected in the reign of Henry VIII.

MR. WEALE'S WORKS ON ARCHITECTURE, ENGINEERING, FINE ARTS, &c.

1 large Engraving, price 3s. 6d.

BRITTON'S (John) VIEWS of the WEST FRONTS of 14 ENGLISH CATHEDRALS.

1 large Engraving in outline, price 2s. 6d.

BRITTON'S (John) PERSPECTIVE VIEWS of the INTERIOR of 14 CATHEDRALS.

In 4to, 2s. 6d.

BRODIE, R., C.E. — Rules for Ranging Railway Curves, with the Theodolite, and without Tables.

1 large Engraving, price 4s. 6d.

BROWN'S (Capt. S.) CHAIN PIER at Brighton, with Details.

The Text in one large volume 8vo, and the Plates, upwards of 70 in number, in an atlas folio volume, very neatly half-bound, £2 10s.

BUCHANAN, R. — PRACTICAL ESSAYS ON MILL WORK AND OTHER MACHINERY; with Examples of Tools of modern Invention; first published by ROBERT BUCHANAN, M.E.; afterwards Improved and edited by THOMAS TREDGOLD, C.E.; and re-edited, with the improvements of the present age, by GEORGE RENNIE, F.R.S., C.E., &c., &c. The whole forming 70 Plates, and 103 Woodcuts.

John Weale, 59, High Holborn, London, W.C.

Text in royal 8vo, and Plates in imperial folio, 18s.

BUCHANAN, R. — SUPPLEMENT. — PRACTICAL EXAMPLES ON MODERN TOOLS AND MACHINES; a Supplementary Volume to Mr. RENNIE'S edition of BUCHANAN "On Mill-Work and Other Machinery," by TREDGOLD. The work consists of 18 Plates.

In 8vo, with Plates, 2nd Edition, 1s 6d.

BURN, C., C.E.—On Tram and Horse Railways.

In one volume, 4to, 21 Plates, half-bound in morocco, £1 1s.

BURY, T., Architect. — Examples of Ancient Ecclesiastical Woodwork.

7s. 6d.

CALCULATOR (THE): Or, TIMBER MERCHANT'S AND BUILDER'S GUIDE. By WILLIAM RICHARDSON and CHARLES GANE, of Wisbeach.

In 8vo, Plates, cloth boards, 7s. 6d.

CALVER, E. K., R.N.—THE CONSERVATION AND IMPROVEMENT OF TIDAL RIVERS.

In 8vo, Woodcuts, 1s 6d.

CALVER, E.K., R.N.—ON THE CONSTRUCTION AND PRINCIPLE OF A WAVE SCREEN, designed for the Formation of Harbours of Refuge.

In 4to, half-bound, price £1 5s.

CARTER, OWEN B., Architect.—A SERIES OF THE ANCIENT PAINTED GLASS OF WINCHESTER CATHEDRAL, Examples of. 28 Coloured Illustrations.

In 4to, 17 Plates, half-bound, 7s. 6d.

CARTER, OWEN B., Architect.—ACCOUNT OF THE CHURCH OF ST. JOHN THE BAPTIST, at Bishopstone, with Illustrations of its Architecture.

John Weale, 59, High Holborn, London, W.C.

MR. WEALE'S WORKS ON ARCHITECTURE, ENGINEERING, FINE ARTS, &c.

Plates, 8vo, boards, 5s.

GWILT, JOSEPH, Architect.—TREATISE ON THE EQUILIBRIUM OF ARCHES.

In 8vo, cloth boards, with 8 Plates, 4s. 6d.

HAKEWELL, S. J.—Elizabethan Architecture; illustrated by parallels of Dorton House, Hatfield, Longle , and Wollaton, in England, and the Palazzo Della Cancellaria at Rome.

8vo, with a Map, 1s.

HAMILTON, P. S., Barrister-at-Law, Halifax Nova Scotia.—Nova Scotia considered as a Field for Emigration.

In imperial 8vo, Third Edition, with additions, 11 Plates, cloth boards, 8s.

HART, J., On Oblique Bridges. — A Practical Treatise on the Construction of Oblique Arches.

In 4to, with Woodcuts, 3s. 6d.

HEALD, GEORGE, C.E.—System of Setting Out Railway Curves.

Royal 8vo, Plates and Woodcuts, price 12s. 6d.

HEDLEY, JOHN. — Practical Treatise on the Working and Ventilation of Coal Mines, with Suggestions for Improvements in Mining.

Two Vols., demy 12mo, in cloth extra boards and lettered, price 12s. 6d.

HOMER. — The Iliad and Odyssey, with the Hymns of Homer, Edition with an accession of English notes by the Rev. T. H. L. LEARY, M.A.

In 8vo, with Engravings, cloth boards, Third Edition, 10s. 6d.

HOPKINSON, JOSEPH, C.E.—The Working of the Steam Engine Explained by the use of the Indicator.

In 8vo, in boards, 18s.

HUNTINGTON, J. B., C.E. — TABLES and RULES for Facilitating the Calculat'on of Earthwork, Land, Curves, Distances, and Gradients, required in the Formation of Railways, Roads, and Canals.

Separate from the above, price 3s.

HUNTINGTON, J. B., C.E. — THE TABLES OF GRADIENTS.

10 Plates, 8vo, bound, 5s.

INIGO JONES.—Designs for Chimney Glasses and Chimney Pieces of the Time of Charles the 1st.

In a sheet, 2s.

IRISH.—Plantation and British Statute Measure (comparative Table of), so that English Measure can be transferred into Irish, and vice versâ.

In 4to, with 8 Engravings, in a wrapper, 6s.

IRON. — ACCOUNT OF THE CONSTRUCTION OF THE IRON ROOF OF THE NEW HOUSES OF PARLIAMENT, with elaborate Engravings of details.

In imperial 4to, with 50 Engravings, and 2 fine Woodcuts, half-bound in morocco, £1 4s.

IRON. — DESIGNS OF ORNAMENTAL GATES, LODGES, PALISADING, AND IRON-WORK OF THE ROYAL PARKS, with some other Designs.

MR. WEALE'S WORKS ON ARCHITECTURE, ENGINEERING, FINE ARTS, &c.

In 4to, with 10 Plates, 12s.

JEBB'S, Colonel, Modern Prisons.—Their Construction and Ventilation.

In 3 vols. 8vo, with 26 elaborate Plates, cloth boards, £2 2s.

JONES, Major-Gen. Sir John, Bart. — Journal of the Sieges carried on by the Army under the Duke of Wellington in Spain, between the years 1811 and 1814, with an Account of the Lines of Torres Vedras. By Major-Gen. Sir JOHN T. JONES, Bart., K.C.B. Third Edition, enlarged and edited by Lieut.-General Sir HARRY D. JONES, Bart.

16mo, cloth boards, 2s. 6d.

KENNEDY AND HACKWOOD'S Tables for Setting out Curves.

In 4to, 37 Plates, half-cloth boards, 9s.

KING, THOMAS.—The Upholsterer's Guide; Rules for Cutting and Forming Draperies, Valances, &c.

Illustrated by large Draughts and Engravings. In 1 volume 4to, text, and a large atlas folio volume of Plates, half-bound, £6 6s.

KNOWLES, JOHN, F.R.S.—The Elements and Practice of Naval Architecture; or, A Treatise on Ship Building, theoretical and practical, on the best principles established in Great Britain; with copious Tables of Dimensions, Scantlings, &c. The Third Edition, with an Appendix, containing the principles of constructing the Royal and Mercantile Navies, by Sir ROBERT SEPPINGS.

41 Plates of a fine and an elaborate description in large atlas folio half-bound, £2 12s. 6d.; with the text half-bound in 4to.

LOCOMOTIVE ENGINES. — The Principles and Practice and Explanation of the Machinery of Locomotive Engines in operation.

In 12mo, sewed, 1s.

MAIN, Rev. ROBERT. — An Account of the Observatories in and about London.

4to, in boards, 15s.

MANUFACTURES AND MACHINERY. — Progress of, in Great Britain, as exhibited chiefly in Chronological notices of some Letters Patent granted for Inventions and Improvements, from the earliest times to the reign of Queen Anne.

16mo, 2s. 6d.

MAY, R. C., C.E.—Method of setting out Railway Curves.

Imperial 4to, with fine Illustrations, extra cloth boards, £1 5s., or half-bound in morocco, £1 11s. 6d.

METHVEN, CAPTAIN ROBERT.—THE LOG OF A MERCHANT OFFICER, Viewed with Reference to the Education of Young Officers and the Youth of the Merchant Service. By ROBERT METHVEN, Commander in the Peninsular and Oriental Company's Service.

In royal 8vo, 1s. 6d.

METHVEN, CAPTAIN ROBERT.—NARRATIVES WRITTEN BY SEA COMMANDERS, ILLUSTRATIVE OF THE LAW OF STORMS. The "Blenheim" Hurricane of 1851, with Diagrams.

Part 1, large 8vo, 5s. Part 2, in preparation.

MURRAY, JOHN, C.E. — A Treatise on the Stability of Retaining Walls, elucidated by Engravings and Diagrams.

MR. WEALE'S WORKS ON ARCHITECTURE, ENGINEERING, FINE ARTS, &c.

50 Plates, 28s., boards.

PORTFOLIO OF GREEK ARCHITECTURE. —Or, Dilettanti Drawing Book; Architectural Engravings, with descriptive Text. Being adapted as studies of the best Classic Models in the Grecian style of Architecture.

50 Plates, £1 8s., bound.

PORTFOLIO OR DRAWING BOOK OF GOTHIC CHURCH ARCHITECTURE.—Of the periods of the 14th, 15th, and 16th centuries. Useful to Architects, Builders, and Students.

25 Plates, folio. 25s.

PORTFOLIO OF ARCHÆOLOGICAL COLLECTIONS.—Of curious, interesting, and ornamental subjects and patterns for stained glass windows, from York.

18 Plates, 10s. 6d. Small folio.

PORTFOLIO OF ANCIENT CAPITAL LETTERS, MONOGRAMS, QUAINT DESIGNS, &c.—Beautifully Coloured and Ornamented.

153 Plates, folio, half-bound in morocco, very neat, £4 4s.

PUBLIC WORKS OF GREAT BRITAIN.—Consisting of Railways, Rails, Chairs, Blocks, Cuttings, Embankments, Tunnels, Oblique Arches, Viaducts, Bridges, Stations, Locomotive Engines, &c.; Cast-Iron Bridges, Iron and Gas Works, Canals, Lock-gates, Centering, Masonry and Brickwork for Canal Tunnels; Canal Boats: the London and Liverpool Docks, Plans and Dimensions, Dock gates, Walls, Quays, and their Masonry; Mooring-Chains; Plan of the Harbour and Port of London, and other important Engineering Works, with Descriptions and Specifications.

In two Parts. Imperial folio.

PUBLIC WORKS OF THE UNITED STATES OF AMERICA.

And the text in an 8vo Volume, price together £2 6s.

REPORTS, SPECIFICATIONS, AND ESTIMATES OF PUBLIC WORKS OF THE UNITED STATES OF AMERICA; explanatory of the Atlas Folio of Detailed Engravings, elucidating practically these important Engineering Works. The Plates are Engraved in the best style.

Imperial 8vo, 50 Engravings, £1 5s.

PAPERS AND PRACTICAL ILLUSTRATIONS OF PUBLIC WORKS OF RECENT CONSTRUCTION—BOTH BRITISH AND AMERICAN. Supplementary to previous Publications, and containing all the details of the Niagara Suspension Bridge.

Half-bound in morocco, finely coloured Plates, price £3 3s.

RAWLINSON'S, ROBERT, C.E.—Designs for Factory, Furnace, and other Tall Chimney Shafts. Tall chimneys are necessary for purposes of Trade and Manufactures. They are required for Factories, for Foundries, for Gas Works, for Chemical Works, for Baths and Wash-houses, and for many other purposes.

Third Edition, in royal 8vo, boards, with 13 Charts, &c., 12s.

REID, Major-General Sir W., F.R.S., &c.—AN ATTEMPT TO DEVELOP THE LAW OF STORMS by means of facts arranged according to place and time; and hence to point out a cause for the variable winds, with a view to practical use in navigation.

John Weale, 59, High Holborn, London, W.C.

MR. WEALE'S WORKS ON ARCHITECTURE, ENGINEERING, FINE ARTS, &c.

In royal 8vo, uniform with the preceding, 9s., with Charts and Woodcuts. The work together in 2 vols., £1 1s.

REID, Major-General Sir W., F.R.S., &c. — THE PROGRESS OF THE DEVELOPMENT OF THE LAW OF STORMS AND OF THE VARIABLE WINDS, with the practicable application of the subject to navigation.

Illustrated with 17 Plates, Third Edition, 8vo, cloth, 7s. 6d.

RICHARDSON, C. J., Architect. — A Popular Treatise on the Warming and Ventilation of Buildings; showing the advantage of the improved system of Heated Water Circulation. And a method to effect the combination of large and small pipes to the same apparatus, and ventilating buildings.

Bound in 2 vols., very neat, half-morocco, gilt tops, price £18.

RENNIE'S, Sir JOHN, F.R.S., Work on the Theory, Formation, and Construction of British and Foreign Harbours, Docks, and Naval Arsenals. This great work may now be had complete, 20 parts and supplement, price £16.

In 8vo, 2s.

RÉVY, J. L., C.E. — THE PROGRESSIVE SCREW AS A PROPELLER IN NAVIGATION.

12mo, cloth boards, 3s. 6d.

SIMMS, F. W. — Treatise on the principal Mathematical and Drawing Instruments employed by the Engineer, Architect, and Surveyor; with a description of the Theodolite, together with Instructions in Field Works.

4to, with fine Plates, a New Edition, extended, sewed, 5s.

SMITH, C. H., Sculptor.—Report and Investigation into the Qualifications and Fitness of Stone for Building Purposes.

In 1 vol. 8vo, in boards, 7s. 6d.

SMITH'S, Colonel of the Madras Engineers, Observations on the Duties and Responsibilities Involved in the Management of Mines.

8vo, cloth boards, with Index Map, 5s.

SOPWITH, THOMAS, F.R.S. — THE AWARD OF THE DEAN FOREST COMMISSIONERS AS TO THE COAL AND IRON MINES.

16 large folio Plates, £1 4s. Separately, 2s. each.

SOPWITH, THOMAS, F.R.S.—SERIES OF ENGRAVED PLANS OF THE COAL AND IRON MINES.

12 Plates; 4to, 6s. in a wrapper.

STAIRCASES, HANDRAILS, BALUSTRADES, AND NEWELS OF THE ELIZABETHAN AGE, &c.—Consisting of — 1. Staircase at Audley-end Old Mauer House, Wilts; 2. Charlton House, Kent; 3. Great Ellingham Hall, Norfolk; 4. Dorfold, Cheshire; 5. Charterhouse; 6. Oak Staircase at Clare Hall, Cambridge; 7. Cromwell Hall, Highgate; 8. Ditto; 9. Catherine Hall, Cambridge; 10. Staircase by Inigo Jones at a house in Chandos Street; 11. Ditto at East Sutton; 12. Ditto, ditto. Useful to those constructing edifices in the early English domestic style.

Large atlas folio Plates, price £2 2s.

STALKARTT, M., N.A. — Naval Architecture; or, The Rudiments and Rules of Ship Building: exemplified in a Series of Draughts and Plans. No text.

John Weale, 59, High Holborn, London, W.C.

MR. WEALE'S WORKS ON ARCHITECTURE, ENGINEERING, FINE ARTS, &c.

With Illustrative Diagrams. In 8vo, 7s. 6d.

STEVENSON'S, THOMAS, C.E., of Edinburgh, Description of the Different kinds of Lighthouse Apparatus.

8vo, 2s. 6d.

STEVENSON, DAVID, C.E., of Edinburgh.—Supplement to his Work on Tidal Rivers.

Text in 4to, and large folio Atlas of 75 Plates, half-cloth boards, £2 12s. 6d.

STEAM NAVIGATION. — Vessels of Iron and Wood; the Steam Engine; and on Screw Propulsion. By WM. FAIRBAIRN, F.R.S., of Manchester; Messrs. FORRESTER, M.I.C.E., of Liverpool; JOHN LAIRD, M.I.C.E., of Birkenhead; OLIVER LANG, (late) of Woolwich; Messrs. SEAWARD, Limehouse, &c. &c. &c. Together with Results of Experiments on the Disturbance of the Compass in Iron-built Ships. By G. B. AIRY, M.A., Astronomer Royal.

10s.

ST. PAUL'S CATHEDRAL, LONDON, SECTION OF. — The Original Splendid Engraving by J. GWYN, J. WALE, decorated agreeably to the original intention of Sir Christopher Wren; a very fine large print, showing distinctly the construction of that magnificent edifice.

Size of Plate 4½ feet in height, 10s.

ST. PAUL'S CATHEDRAL, LONDON, GREAT PLAN. — J. WALE and J. GWYN'S GREAT PLAN, accurately measured from the Building, with all the Dimensions figured and in detail, description of Compartments by engraved Writing.

Second Edition, greatly enlarged, royal 8vo, with Plates, cloth boards, price 16s.

STRENGTH OF MATERIALS.—FAIRBAIRN, WILLIAM, C.E., F.R.S., and of the Legion of Honour of France. On the application of Cast and Wrought Iron to Building Purposes.

With Plates and Diagrams. New Edition. The work complete in 2 vols., bound in 1 vol., price, in cloth boards, 16s. The second portion of the work, containing Mr. Hodgkinson's Experimental Researches, may be had separately, price 9s.

STRENGTH OF MATERIALS.—HODGKINSON, EATON, F.R.S., AND THOMAS TREDGOLD, C.E. A PRACTICAL ESSAY ON THE STRENGTH OF CAST IRON AND OTHER METALS; intended for the assistance of engineers, ironmasters, millwrights, architects, founders, smiths and others engaged in the construction of machines, buildings, &c' By EATON HODGKINSON, F.R.S.

To be published in 1861, in crown 8vo, bound for use.

STRENGTH OF MATERIALS.—POLE, WILLIAM, C.E., F.R.S.—Tables and popular explanations of the Strength of Materials, of Wrought and Cast Iron with other metals, for structural purposes; devèloping in a systematic form, the strengths, bearings, weights, and forms of these materials, whether used as girders or arches, for the construction of bridges and viaducts, public buildings, domestic mansions, private buildings, columns or pillars, bressummers for warehouses, shops, working and manufacturing factories, &c. &c. &c. The whole rendered of easy reference for architects, builders, civil and mechanical engineers, millwrights, ironfounders, &c. &c. &c., and forming Ready Reckoner or Calculator.

John Weale, 59, High Holborn, London, W.C.

MR. WEALE'S WORKS ON ARCHITECTURE, ENGINEERING, FINE ARTS, &c.

30 very elaborately drawn Engravings. In large 4to, neatly half-bound and lettered, £1 1s. A few copies on large imperial size, extra half-binding, £1 11s. 6d.

TEMPLE CHURCH.—The Architectural History and Architectural Ornaments, Embellishments, and Painted Glass, of the Temple Church, London.

Part I., with 28 Engravings on Wood and Copper, in cloth boards, 4to, 15s.

THAMES TUNNEL.—A Memoir of the several Operations and the Construction of the Thames Tunnel, from Papers by the late Sir ISAMBARD BRUNEL, F.R.S., Civil Engineer.

Fourth Edition, with a Supplementary Addition, large 8vo, 12s. 6d.

THOMAS (LYNALL), F.R.S.L.—Rifled Ordnance. —A Practical Treatise on the Application of the Principle of the Rifle to Guns and Mortars of every calibre; to which is added a New Theory of the Initial Action and Force of Fixed Gunpowder plates.

In 4to, complete, cloth, Vol. I., with Engravings, £1 10s.; Vol. II., ditto, £1 8s.; Vol. III., ditto, £2 12s. 6d.

TRANSACTIONS OF THE INSTITUTION OF CIVIL ENGINEERS.

8 vols., numerous Engravings of Sections of Coal Mines, &c., large folding Plates, several of which are coloured, in large 8vo, half-bound in calf, price £1 1s. per volume.

TRANSACTIONS OF THE NORTH OF ENGLAND INSTITUTE OF MINING ENGINEERS.—Commencing in 1852, and continued to 1860.

A New Edition revised by the translator, and with additional Plates, in demy 12mo, India proof Plates and Vignettes, half-bound in morocco, gilt tops, price 12s. Only 25 printed on India paper.

VITRUVIUS. — The Architecture of Marcus Vitruvius Pollio in 10 Books. Translated from the Latin by JOSEPH GWILT, F.S.A., F.R.A.S.

In 4to, with Plates, 7s. 6d.

WALKER'S, THOMAS, Architect. — Account of the Church at Stoke Golding.

£1 10s.

WEALE'S QUARTERLY PAPERS ON ENGINEERING. — Vol. VI. (Parts 11 and 12 completing the work.) Comprising, "On the Principles of Water Power." Plates. Experiments on Locomotive Engines. Coloured Plates. On Naval Arsenals. On the Mode of Forming Foundations under water and on bad ground. Plates. On the Improvement of the River Medway and of the Port and Arsenal of Chatham. On the Improvement of Portsmouth Harbour. An Analysis of the Cornish Pumping. Plates. On Water Wheels. Plates.

Text in 8vo, cloth boards, and Plates in atlas folio, in cloth, 16s.

WHITE'S, THOMAS, N.A., Theory and Practice of Ship Building.

In 8vo, with a large Sectional Plate, 1s. 6d.

WHICHCORD, JOHN, Architect. — OBSERVATIONS ON KENTISH RAG STONE AS A BUILDING MATERIAL.

John Weale, 59, High Holborn, London, W.C.

MR. WEALE'S WORKS ON ARCHITECTURE, ENGINEERING FINE ARTS, &c.

4to, coloured Plates, in half-morocco, 7s 6d.

WHICHCORD, JOHN, Architect.—HISTORY AND ANTIQUITIES OF THE COLLEGIATE CHURCH OF ALL SAINTS, MAIDSTONE.

In 4to, 6s.

WICKSTEED, THOMAS, C.E. — AN EXPERIMENTAL INQUIRY CONCERNING THE RELATIVE POWER OF, AND USEFUL EFFECT PRODUCED BY, THE CORNISH AND BOULTON & WATT PUMPING ENGINES, and Cylindrical and Waggon-Head Boilers.

In 8vo, 1s.

WICKSTEED, THOMAS, C.E. — FURTHER ELUCIDATION OF THE USEFUL EFFECTS OF CORNISH PUMPING ENGINES; showing the average working for long periods, &c., &c., &c.

£2 2s.

WICKSTEED, THOMAS, C.E. — THE ELABORATELY ENGRAVED ILLUSTRATIONS OF THE CORNISH AND BOULTON & WATT ENGINES erected at the East London Water Works, Old Ford. Eight large atlas folio very fine line engravings by GLADWIN, from elaborate drawings made expressly by Mr. WICKSTEED; folio, together with a 4to explanation of the plates, containing an engraving, by LOWRY, of Harvey and West's patent pump-valve, with specification.

With numerous Woodcuts.

WILLIAMS, C. WYE, Esq., M. Inst. C. E.—THE COMBUSTION OF COAL AND THE PREVENTION OF SMOKE, chemically and practically considered.

Imperial 8vo, with a Portrait, 2s. 6d.

WILLIAMS, C. WYE, Esq., M. Inst. C. E. — PRIZE ESSAY ON THE PREVENTION OF THE SMOKE NUISANCE, with a fine portrait of the Author.

With 3 Plates, containing 51 figures, 4to, 5s.

WILLIS, REV. PROFESSOR, M.A.—A system of Apparatus for the use of Lecturers and Experimenters in Mechanical Philosophy.

In 4to, bound, with 26 large plates and 17 woodcuts, 12s.

WILME'S MANUALS. — A MANUAL OF WRITING AND PRINTING CHARACTERS, both ancient and modern.

Maps and Plans, in 4to, plates coloured, half-bound morocco, £2.

WILME'S MANUALS. — A HANDBOOK FOR MAPPING, ENGINEERING, AND ARCHITECTURAL DRAWING.

Three Vols., large 8vo, £3.

WOOLWICH. — COURSE OF MATHEMATICS. This course is essential to all Students destined for the Royal Military Academy at Woolwich.

8vo, 1s.

YULE, MAJOR-GENERAL.—ON BREAKWATERS AND BUOYS of VERTICAL FLOATS.

John Weale, 59, High Holborn, London, W.C.

FOREIGN WORKS, KEPT IN STOCK AS FOLLOWS:—

Large folio, 32 plates, some coloured, and 12 woodcuts, 50 francs. £2 10s.

ARCHITECTURE SUISSE.—Ou Choix de Maisons Rustiques des Alpes du Canton de Berne, par GRAFFINRIED et STÜRLER, Architectes. Berne, 1844.

Small folio, 52 most interesting and explanatory plates of Public Works, Bridges, Iron Works, &c., &c., &c., very neatly half-bound in morocco, £1 10s.

BAUERNFEIND, CARL MAX.—VORLEGE-BLAETTER ZUR BRUCKENBAU KUNDE. München.

Large folio, 36 plates of Byzantine capitals, 12s.

BYZANTINISCHE CAPITAELER.—München.

Second edition, 126 plates, large folio, best Paris edition, 100 f., printed on fine paper, half-cloth boards, £4 4s.

CALLIAT, VICTOR, ARCT.—Parallèle des Maisons de Paris, construites depuis 1830 jusqu'à nos jours.—1857.

Large folio, 60 francs, 60 plates, and several vignettes, £2 8s.

CANÉTO, F.—Sainte-Marié d'Auch. Atlas Monographique de Cette Cathédrale. The Plates consist principally of outline drawings of the Painted Glass Windows in this Cathedral.

120 plates, elegant in half-morocco extra, interleaved, £5 15s. 6d.

CASTERMAN, A.—PARALÈLLE des MAISONS de BRUXELLES et des PRINCIPALES VILLES de la BELGIQUE, construites depuis 1830 jusqu'à nos jours, représentés en plans, élévations, coupes et détails intérieurs et extérieurs. —Paris.

Small folio, 48 plates of edifices, £1 1s.

DEGEN, L.—LES CONSTRUCTIONS EN BRIQUES, composées et publiées. 8 livraisons.—1858.

Small folio, 48 plates of houses, parts of houses, details of all kinds of singularly beautiful woodwork, coloured plates in imitation of the objects given, £1 1s.

DEGEN, L.—LES CONSTRUCTIONS ORNA-MENTALES EN BOIS, 8 livraisons.

In 3 very large folio parts, 85 fine plates, £1 11s. 6d.

GAERTNER, F. V.—The splendid works of M. GAERTNER of Munich, drawn to a very large size, consisting of the library in plans, elevations, interiors, details, and sections, and coloured ornaments. The church, with details, ornaments, &c.—München.

Small folio, 86 fine plates of the Architecture, ornament, and detail of the houses and churches of Germany during the middle age, very neatly half-bound in morocco, £2 12s. 6d.

KALLENBACH, C. C.—Chronologie der Deutsch-Mittelaltorlichen Baukunst.—München. Fine Work.

The works of the great master KLENZIE of Munich, in 5 parts very large folio, 50 plates of elevations, plans, sections, details and ornaments of his public and private buildings executed in Munich and St. Petersburg, £2 2s.

KLENZE, LEO VON.—Sammlung Arthitec-tonisher Entwürfe, für die Ausführung bestimmt oder wirklich ausgeführt. Published in Munich.

FOREIGN WORKS KEPT IN STOCK AS FOLLOWS.

Large folio, containing a profusion of Plates of the palaces, theatres, hôtel de villes, and other public buildings in several parts of Italy. Elegantly half-bound in red morocco, extra gilt and interleaved, £8 6s.

FABBRICHE.—E DISEGNI D'ANTONIO DIEDO, NOBILE VENETO. Venezia.

86 livraisons, price £12 12s.

GALLERIA DI TORINO (LA REALE).—Illustrata da R. D'AZEGLIO, Memb. dell' Accad., &c. &c. Copies, India proofs, £18 18s.

*** Bound copies in elegant half-morocco binding. India proof, £23 2s.

2 vols. folio, complete, 177 Plates of outline elevations, plans, interiors, details, &c., first impression, 150 francs, half-bound. £6 6s.

GAUTHIER, M.P., Architecte.—Les PLUS BEAUX ÉDIFICES de la VILLE de GENES et des ses ENVIRONS. Paris, 1830-2.

Folio, 109 Plates of plans, elevations, sections, and details, £2 8s.

GRANDJEAN de MONTIGNY et A. FAMIN.—ARCHITECTURE TOSCANE, ou palais, maisons, et autres édifices, de la Toscane. Paris, 1815.

Oblong folio, containing a profusion of picturesque views of palaces and public buildings and scenes of Venice, executed in tinted lithography, with full descriptions attached to each. Elegant in half extra morocco, interleaved, £4 14s. 6d.

KIER, G.—VENEZIA MONUMENTALE PITTORESCA. Venezia.

Large folio, 61 livraisons or 3 vols., with 3 vols. of text in 4to, £18 18s.

LETAROUILLY, P.—Édifices de Rome Moderne. Paris, 1825-55.

Fine Plates of the New Palace of Justice, Senate House, &c., plans, elevations, sections, doors, &c., details of the several parts, &c., £1 1s.

MICHELA, IGNAZIO.—DESCRIZIONE e DISEGNI del PALAZZO dei MAGISTRATI SUPREMI di TORINO. Torino.

Large folio, 94 Plates, bound in extra half-morocco, gilt and interleaved, price £6 10s.

REYNAUD, L.—Trattato di Architettura, contenente nozioni generali sui Principii della Construzione e sulla storia dell' Arti, con annot. per cura di Lorenzo Urbani. Venezia, 1857.

4 imperial bulky 8vo volumes, printed and published under authority, and treats of the early foundation of Venice and establishment as a kingdom, its wealth and commerce, and its once great political position, with Plates, £3 3s.

VENEZIA.—E le sue Lagune. Venezia, 1847.

VENEZIA.—Copies elegantly bound and gilt, £4 14s. 6d. Venezia, 1847.

In 2 large folio volumes, numerously and elaborately drawn Plates, very well executed in outline, altogether a very fine work. Very elegantly half-bound in morocco, extra gilt and interleaved, £12 12s.

ACCADEMIA DI BELLI ARTI.—Opere dei Grandi Concorsi Premiate dall' I. R. Accademia delle Belle; Arti, in Milano, e publicate, per cura dell' Architetto, G. ALUISETTI—per la Classi di Ornato—per le Classi di Architettura, figura ed Ornato. Milano, 1825-29.

John Weale, 59, High Holborn, London, W.C.

FOREIGN WORKS, KEPT IN STOCK AS FOLLOWS:—

Atlas folio, very fine impressions, complete in 3 parts, Columbier folio, £3 13s. 6d Elegantly half-bound in extra morocco and interleaved, £5 15s. 6d.

ALBERTOLLI, G.—Alcune Decorazioni di Nobili Sale ed Altri Ornamenti. Milano, 1787, 1824, 1838.

To be had separately, £1 8s.

ALBERTOLLI, G.—Part III., very frequently required to make up sets.

---

2 vols., folio, 80 Plates of the most exquisite kind in colours, far superior to any existing work of the present day, £7 10s.

HOFFMAN, ET KELLERHOVEN. — Recueil de Dessins relatifs à l'Art de la Décoration chez tous les peuples et aux plus belles époques de leur civilisation, &c., destinés à servir de motifs et de matériaux aux peintres, décorateurs, peintres sur verre, et aux dessinateurs de fabriques.

Price £1 1s.

HOPE, ALEXANDER J. BERESFORD, Esq.— Abbildungen der Glasgemälde in der Salvator-Kirche zu Kilndown in der Graffschaft Kent. Copies of paintings on glass in Christ Church, Kilndown, in the county of Kent, executed in the Royal Establishment for Painting on Glass, Munich, by order of ALEXANDER J. BERESFORD HOPE, Esq., published by F. Eggert, Painter on Glass, München. The work contains one sheet with the dedication to A. J. B. HOPE, Esq., and fourteen windows; in the whole fifteen, beautifully engraved and carefully coloured.

In large folio, 80 Plates, containing a profusion of rich Italian and other ornaments. Elegant in half-morocco, gilt, and interleaved, £6 6s.

JULIENNE, E.—Industria Artistica o Raccolto di Composizioni e Decorazioni Ornamentali, come suppellettili, tappezzerie, armature, cristalli, soffitti, cornici, lampade, bronzi, ec. Venezia, 1851—1856.

Prix 50f., in folio, £3.

LE PAUTRE.—Collection des plus belles Compositions, gravées par DE CLOUX, Archte. L'Ouvrage contient cent planches. Paris.

This unique collection is in 2 Vols. 4to, had its commencement in 1812, and contains upwards of 500 rich Designs. Price £5 5s.

METIVIER, MONS., Architecte.—The original Sketches, Drawings, and Tracings, in pencil and pen and ink, of executed Works and Proposals, displaying the genius of Mons. Metivier, as an architect of high attainments, whose recent death was much regretted in Bavaria. He was a native of France, and was induced to settle in Munich by the late Duke of Leuchtenberg, under whose patronage he was much employed in the construction of private edifices for the Bavarian nobility and gentry; and for decoration and fittings of them; his interiors are still much in admiration. He built a mansion for Prince Charles, in a most simple and elegant style (in Brienner Street), which is still now considered one of the purest buildings of Munich. The above Sketches are his professional life and practice.

Twelve Parts, in small oblong 4to, 60 coloured Plates of 90 elaborately coloured and gilt ornaments. £1 1s.

ORNAMENTENBUCH.—Farbige Verzierungen für Fabrikanten, Zimmermaler und andere Baugewerke. München.

John Weale, 59, High Holborn, London, W.C.

FOREIGN WORKS, KEPT IN STOCK AS FOLLOWS;—

410 Plates, in two thick large 4to. Vols., designed and engraved by MM. Reister Arget, d'Hautel, de Wailly, Wagner, L. Feuchère et Regnier, &c. £5 5s.

ORNEMENTS.—Tirés ou imités des Quatre Écoles. Paris.

Six Parts, large folio, Plates beautifully coloured, in fac-similes of the Interiors, Ornaments, Compartments, Ceilings, &c. £2 12s. 6d. Also, elegantly half-bound in morocco gilt, £4 4s.

ROTTMANN, L.—Ornamente aus den vorzüglichsten Bauwerken Münchens. München.

Very elegant, in half red morocco, gilt, and interleaved, £7 17s. 6d.

ZANETTI, G.—STUDII ARCHITETTONICO ORNAMENTALI, dedicati all' J. R. Accademia Veneta delle Belle Arti, seconda edizione con aggiunte del Prof. L. URBANI. 56 livraisons, in imperial folio, about 200 of most elaborately designed subjects of Architecture and Interior Fittings, Designs for Chimney Pieces, Iron Work for Interiors and Exteriors, Gates and Wooden Gates, Garden Decorations, &c., &c., including the Appendices. Venezia.

A Catalogue, of 40 pages, to be had gratis; printed in demy 8vo.

Export Orders executed either for Principals abroad, or Merchants at home.

www.ingramcontent.com/pod-product-compliance
Lightning Source LLC
Chambersburg PA
CBHW021705210326
41599CB00013B/1532